Chapter 1: Introduction to AI Agents

Artificial intelligence (AI) has transformed numerous aspects of our daily lives, from how we communicate and work to how we interact with technology. At the heart of this revolution are AI agents—intelligent entities designed to perceive their environment, reason, and take action autonomously or semi-autonomously. This chapter introduces the concept of AI agents, explores the different types, and provides an overview of their historical evolution.

Definition of AI Agents

An AI agent is an entity that perceives its environment through sensors and acts upon that environment through actuators. In simpler terms, an AI agent can be thought of as a program or a system that can make decisions based on input from its surroundings. The actions taken by an AI agent can range from simple responses to complex decision-making processes, depending on the agent's design and purpose.

Key Characteristics of AI Agents

1. **Autonomy**: AI agents can operate independently, making decisions without human intervention. This autonomy allows them to respond to dynamic environments in real-time.
2. **Perception**: AI agents use various sensors to gather information about their surroundings. This can include data from cameras, microphones, or other input devices.
3. **Action**: Once an AI agent processes its perceptions, it can take action to achieve specific goals. This could involve physical actions (like moving a robot) or digital actions (like sending an email).
4. **Learning**: Many AI agents incorporate machine learning techniques that allow them to improve their performance over time based on experience and new data.

Types of AI Agents

AI agents can be categorized into several types based on their functionality and operational characteristics. Understanding these types helps in grasping the diverse applications and implications of AI technology.

1. Reactive Agents

Reactive agents operate based on immediate input from their environment without maintaining an internal model of the world. They respond to stimuli directly, making them suitable for simple tasks. For example, a basic chatbot that provides answers based on keyword recognition can be classified as a reactive agent.

2. Proactive Agents

Proactive agents not only respond to their environment but also anticipate future needs or events. They can maintain an internal model of the environment and plan their actions accordingly. For instance, a virtual assistant that schedules meetings based on user preferences and availability is a proactive agent.

3. Autonomous Agents

Autonomous agents function independently, capable of making complex decisions and executing tasks without human intervention. They often use advanced AI techniques, including deep learning and reinforcement learning. Self-driving cars are a prime example of autonomous agents, as they navigate and make decisions in real-time based on their surroundings.

4. Multi-Agent Systems

In multi-agent systems, multiple AI agents interact with one another to achieve common goals. These systems can be used to solve complex problems that require collaboration among various agents, such as optimizing logistics in supply chain management or coordinating efforts in robotics.

Historical Evolution of AI Agents

The concept of AI agents has evolved significantly over the decades, shaped by advancements in technology, theories of intelligence, and changing societal needs.

1. Early Concepts and Theoretical Foundations

The idea of artificial agents can be traced back to the mid-20th century, with foundational work in computer science and mathematics. Alan Turing's seminal paper on machine intelligence in 1950 laid the groundwork for future AI developments. Early AI research focused on problem-solving and symbolic reasoning, leading to the creation of simple agents that could perform tasks based on predefined rules.

2. The Rise of Expert Systems

In the 1980s, expert systems emerged as a prominent type of AI agent designed to mimic human expertise in specific domains. These systems relied on large rule-based databases to make decisions and provide recommendations, finding applications in fields like medicine, finance, and engineering. However, their rigid structures limited adaptability.

3. Machine Learning Revolution

The late 1990s and early 2000s saw a resurgence in AI research driven by advancements in machine learning and increased computational power. Agents began to incorporate learning algorithms that allowed them to adapt and improve over time. This era marked the transition from rule-based systems to data-driven approaches, enhancing the capabilities of AI agents.

4. Current Trends and Future Directions

Today, AI agents leverage deep learning, natural language processing, and other advanced technologies to operate in increasingly complex environments. The rise of conversational agents, autonomous vehicles, and robotic systems exemplifies the ongoing evolution of AI agents. As AI continues to advance, the future promises even more sophisticated agents capable of understanding and interacting with the world in human-like ways.

Conclusion

AI agents represent a fascinating convergence of technology, intelligence, and practical application. By understanding their definitions, types, and historical evolution, we can appreciate the potential of these intelligent entities in shaping our future. As we move forward in this book, we will explore the fundamentals of artificial intelligence, the design principles of AI agents, and their diverse applications across various fields. Mastering AI agents not only enhances our technological capabilities but also raises important ethical and societal questions that we must address as we continue this journey.

Chapter 2: The Fundamentals of Artificial Intelligence

Artificial Intelligence (AI) is a multidisciplinary field that encompasses a variety of technologies and methodologies aimed at creating machines capable of intelligent behavior. To master AI agents, it is essential to understand the core concepts of AI, the distinctions between machine learning and traditional programming, and the various technologies that underpin modern AI applications. This chapter will provide an overview of these fundamental aspects, setting the stage for deeper exploration in subsequent chapters.

Core Concepts of AI

1. Definition of AI

AI can be defined as the simulation of human intelligence processes by machines, particularly computer systems. These processes include learning (the acquisition of information and rules for using it), reasoning (using rules to reach approximate or definite conclusions), and self-correction. The ultimate objective of AI is to create systems that can perform tasks that typically require human intelligence.

2. Categories of AI

AI can be categorized into two primary types:

- **Narrow AI**: Also known as weak AI, narrow AI refers to systems designed to handle a specific task or a narrow range of tasks. Examples include virtual assistants like Siri and chatbots that can manage customer inquiries.
- **General AI**: General AI, or strong AI, is a theoretical form of AI that possesses the ability to understand, learn, and apply intelligence across a broad range of tasks, akin to human cognitive abilities. As of now, general AI remains largely speculative and is a subject of ongoing research.

3. Key Components of AI

The main components of AI include:

- **Machine Learning**: A subset of AI that involves training algorithms to learn from and make predictions based on data. Machine learning enables systems to improve their performance over time without being explicitly programmed for every task.
- **Natural Language Processing (NLP)**: The branch of AI that focuses on the interaction between computers and humans through natural language. NLP enables machines to understand, interpret, and generate human language in a valuable way.
- **Computer Vision**: This field enables machines to interpret and understand visual information from the world, simulating human sight. Computer vision is used in applications like facial recognition, autonomous vehicles, and image analysis.

Machine Learning vs. Traditional Programming

One of the fundamental shifts in AI development is the move from traditional programming to machine learning. Understanding this distinction is crucial for grasping how AI agents operate.

1. Traditional Programming

In traditional programming, a human programmer writes explicit instructions for a computer to follow. The logic is defined step by step, and the output is predictable based on the input provided. For example, a calculator follows a set of predefined rules to perform arithmetic operations.

2. Machine Learning

In contrast, machine learning does not rely on explicit programming. Instead, it allows systems to learn from data. The process generally involves:

- **Training**: The system is fed a large dataset, from which it learns patterns and relationships. For instance, a machine learning model for image recognition may be trained on thousands of images labeled with their respective categories.
- **Testing**: After training, the model is tested on new, unseen data to evaluate its performance. This step helps determine how well the model generalizes to new situations.
- **Improvement**: The model can be iteratively improved by retraining it with additional data or by fine-tuning its parameters based on performance metrics.

Machine learning allows for more flexibility and adaptability, enabling AI agents to function effectively in dynamic environments.

Overview of AI Technologies

1. Neural Networks

Neural networks are a key technology in AI, inspired by the structure and functioning of the human brain. These networks consist of interconnected nodes (neurons) that process information in layers. Neural networks are particularly effective for tasks such as image recognition and natural language processing, thanks to their ability to learn complex patterns from large datasets.

2. Natural Language Processing (NLP)

NLP technologies enable machines to understand, interpret, and respond to human language. Key techniques in NLP include:

- **Tokenization**: Breaking down text into smaller components, such as words or phrases, to analyze meaning.
- **Sentiment Analysis**: Determining the emotional tone behind a series of words, useful for applications like social media monitoring.
- **Language Generation**: Producing coherent and contextually relevant text based on input data, enabling the creation of chatbots and virtual assistants.

3. Computer Vision

Computer vision technologies enable machines to interpret visual data. This involves:

- **Image Recognition**: Identifying objects, people, or scenes within images.
- **Object Detection**: Locating and classifying multiple objects within an image, crucial for applications like autonomous vehicles.
- **Facial Recognition**: Identifying and verifying individuals based on facial features, widely used in security systems.

4. Robotics

Robotics integrates AI with mechanical systems to create machines that can perform tasks autonomously. Key areas of focus include:

- **Robotic Process Automation (RPA)**: Using software robots to automate repetitive tasks in business processes.
- **Autonomous Robots**: Robots equipped with AI agents that can navigate and make decisions in dynamic environments, such as drones and self-driving cars.

Conclusion

Understanding the fundamentals of artificial intelligence is essential for mastering AI agents. By grasping the core concepts, recognizing the differences between machine learning and traditional programming, and becoming familiar with the technologies that drive AI, we can appreciate the potential and challenges of AI agents in various applications. As we progress in this book, we will delve deeper into designing AI agents, exploring their types, and understanding the technologies that enhance their capabilities. The journey into the world of AI agents is just beginning, and the possibilities are vast.

Chapter 3: Designing AI Agents

The design of AI agents is a critical aspect that influences their functionality, adaptability, and overall effectiveness. A well-designed AI agent can seamlessly integrate into various applications, enhancing user experiences and achieving specific goals. This chapter explores the key principles of agent design, understanding the environments in which they operate, and the tools and frameworks that support their development.

Key Principles of Agent Design

1. Modularity

Definition: Modularity refers to the design principle where an AI agent is constructed from independent, interchangeable components or modules. Each module handles specific tasks or functionalities.

Benefits:

- **Flexibility**: Modular agents can be easily updated or replaced without overhauling the entire system.
- **Reusability**: Components can be reused across different agents or applications, saving time and resources.

2. Autonomy

Definition: Autonomy is the ability of an AI agent to operate independently and make decisions without human intervention.

Benefits:

- **Efficiency**: Autonomous agents can perform tasks in real-time, reducing the need for constant oversight.
- **Adaptability**: They can learn from their experiences and adapt to new situations, enhancing their effectiveness.

3. Interactivity

Definition: Interactive agents can communicate and collaborate with users and other agents. They should be designed to understand and respond to user inputs effectively.

Benefits:

- **User Engagement**: High interactivity fosters better user experiences and increases satisfaction.
- **Collaboration**: Interactive agents can work together to solve complex problems, leveraging each other's strengths.

4. Learning and Adaptation

Definition: Incorporating learning capabilities allows AI agents to improve their performance over time by adapting to new data and experiences.

Benefits:

- **Improved Accuracy**: Agents can refine their algorithms and strategies based on past performance, leading to better decision-making.
- **Dynamic Responses**: Learning agents can adjust their behavior in response to changing environments, making them more effective.

5. Robustness

Definition: Robustness refers to the ability of an AI agent to function correctly under a variety of conditions, including unexpected inputs or environmental changes.

Benefits:

- **Reliability**: Robust agents can maintain performance even in the face of challenges, ensuring consistent outcomes.
- **Safety**: Ensuring that agents can handle failures gracefully reduces the risk of harmful consequences.

Understanding Agent Environments

AI agents operate within specific environments that influence their behavior and decision-making processes. Understanding these environments is essential for effective agent design.

1. Static vs. Dynamic Environments

- **Static Environments**: These are environments that do not change while the agent is operating. For example, a chess game has a static environment where the rules and state of the game remain constant until a move is made.
- **Dynamic Environments**: In these environments, conditions can change unpredictably, requiring agents to adapt their strategies in real-time. Examples include self-driving cars navigating city streets or stock trading agents responding to market fluctuations.

2. Fully Observable vs. Partially Observable Environments

- **Fully Observable**: The agent has access to all the relevant information about its environment. An example is a game of tic-tac-toe, where the agent can see the entire board.
- **Partially Observable**: The agent has limited access to information, which complicates decision-making. For instance, in poker, an agent cannot see the opponents' cards and must make decisions based on incomplete information.

3. Discrete vs. Continuous Environments

- **Discrete Environments**: These have a finite number of distinct states and actions. Board games are examples of discrete environments, where moves are limited and well-defined.
- **Continuous Environments**: In contrast, continuous environments have an infinite range of states and actions, such as navigating a robot in a physical space where movement can vary in many dimensions.

Tools and Frameworks for Developing AI Agents

The development of AI agents requires a combination of programming languages, libraries, and frameworks tailored for specific tasks and environments. Here are some essential tools and frameworks:

1. Programming Languages

- **Python**: Widely used for AI development due to its simplicity and extensive libraries, including TensorFlow, Keras, and PyTorch for machine learning.
- **Java**: Known for its portability and scalability, Java is often used in enterprise-level AI applications.

2. Machine Learning Libraries

- **TensorFlow**: An open-source library for machine learning and deep learning, widely used for developing neural networks and other AI models.
- **Scikit-learn**: A Python library that provides simple and efficient tools for data mining and data analysis, particularly for traditional machine learning algorithms.

3. Natural Language Processing Frameworks

- **NLTK (Natural Language Toolkit)**: A comprehensive library for natural language processing in Python, providing tools for tasks such as tokenization, parsing, and sentiment analysis.
- **spaCy**: An efficient library for NLP, designed for production use. It is fast and easy to integrate into applications requiring text processing.

4. Robotics Frameworks

- **ROS (Robot Operating System)**: A flexible framework for writing robot software. It provides tools and libraries to help developers create robot applications.
- **OpenAI Gym**: A toolkit for developing and comparing reinforcement learning algorithms, providing various environments for testing and training AI agents.

Conclusion

Designing effective AI agents involves a careful consideration of key principles, understanding the environments in which they operate, and leveraging the appropriate tools and frameworks. By focusing on modularity, autonomy, interactivity, learning, and robustness, developers can create agents that not only perform tasks effectively but also adapt to changing conditions and user needs. As we progress through this book, we will explore the various types of AI agents, delve deeper into the concepts of machine learning, and examine the technologies that power these intelligent systems. The design and development of AI agents represent a fascinating intersection of technology and creativity, paving the way for innovative applications that can significantly impact our lives.

Chapter 4: Types of AI Agents

In the realm of artificial intelligence, agents can be classified into various categories based on their functionalities, capabilities, and applications. Understanding these different types of AI agents is crucial for mastering their design, implementation, and use in real-world scenarios. This chapter delves into the primary classifications of AI agents, including autonomous vs. non-autonomous agents, conversational agents (like chatbots and virtual assistants), and robotic agents, along with their respective applications.

Autonomous vs. Non-Autonomous Agents

1. Autonomous Agents

Definition: Autonomous agents operate independently in their environments, making decisions without human intervention. They are capable of perceiving their surroundings, reasoning based on available information, and taking actions to achieve specific goals.

Characteristics:

- **Self-Directed**: Autonomous agents can initiate actions based on their internal goals or external stimuli.
- **Adaptability**: They can learn from experiences and adjust their behavior over time, improving their performance and decision-making abilities.

Examples:

- **Self-Driving Cars**: These vehicles utilize a combination of sensors, machine learning algorithms, and advanced decision-making processes to navigate complex environments safely.
- **Drones**: Autonomous drones can perform tasks such as aerial surveillance or package delivery by processing real-time data and making navigation decisions.

2. Non-Autonomous Agents

Definition: Non-autonomous agents require human input or intervention to function. They can be reactive or operate based on predefined rules, but they do not possess the capability to act independently.

Characteristics:

- **Dependent on Users**: Non-autonomous agents often rely on human operators for guidance and decision-making.
- **Limited Learning**: While they may incorporate some basic learning mechanisms, they cannot adapt as flexibly as autonomous agents.

Examples:

- **Simple Chatbots**: Basic chatbots that respond to predefined keywords or phrases without understanding context are non-autonomous agents.
- **Industrial Robots with Human Control**: These robots perform tasks based on direct commands from operators, lacking independent decision-making capabilities.

Conversational Agents

Conversational agents are designed to engage in dialogue with users, simulating human-like interactions. They can be categorized into two main types: chatbots and virtual assistants.

1. Chatbots

Definition: Chatbots are AI programs that simulate conversation through text or voice interactions. They can be rule-based (following a fixed set of responses) or use machine learning to improve their conversational abilities.

Applications:

- **Customer Service**: Many businesses deploy chatbots on their websites to handle common inquiries, providing instant responses to customer questions.
- **Social Media**: Chatbots can engage users on platforms like Facebook Messenger, offering personalized interactions and facilitating communication.

Example:

Zendesk Chat

2. Virtual Assistants

Definition: Virtual assistants are more advanced conversational agents capable of performing a wider range of tasks and providing personalized assistance based on user preferences and context.

Applications:

- **Personal Productivity**: Virtual assistants like Amazon's Alexa, Apple's Siri, and Google Assistant help users manage schedules, set reminders, and control smart home devices.
- **Information Retrieval**: They can answer questions, provide recommendations, and access various online services, making them versatile tools for daily tasks.

Example:

Google Assistant

Robotic Agents

Robotic agents are physical entities equipped with AI capabilities, enabling them to perform tasks in the real world. They often integrate various technologies, including computer vision, sensors, and machine learning.

1. Types of Robotic Agents

- **Industrial Robots**: Used in manufacturing settings for tasks such as assembly, welding, and quality control. They can operate autonomously or be guided by human operators.
- **Service Robots**: Designed to assist with everyday tasks in settings like hospitals, restaurants, and households. Examples include robotic vacuum cleaners and robotic waiters.

2. Applications of Robotic Agents

- **Healthcare**: Surgical robots assist surgeons with precision, while rehabilitation robots aid patients in recovery.
- **Exploration**: Robotic agents are employed in environments that are dangerous or inaccessible to humans, such as space exploration or underwater research.

Example:

Boston Dynamics' Spot

Conclusion

Understanding the types of AI agents is essential for effectively applying them in real-world scenarios. Autonomous agents, non-autonomous agents, conversational agents, and robotic agents each serve unique roles, equipped with distinct capabilities and applications. As we continue our journey through this book, we will delve deeper into the mechanisms and technologies that underpin these agents, exploring machine learning, natural language processing, and the broader implications of AI in various domains. By mastering these concepts, we can harness the potential of AI agents to solve complex problems and enhance our everyday lives.

Chapter 5: Understanding Machine Learning

Machine learning (ML) is a pivotal component of artificial intelligence that empowers AI agents to learn from data and improve their performance over time. By distinguishing between different types of machine learning, understanding data preparation and feature engineering, and exploring various algorithms, we can appreciate how machine learning underpins the functionality of AI agents. This chapter delves into these essential aspects of machine learning.

Types of Machine Learning

Machine learning is broadly categorized into three main types: supervised learning, unsupervised learning, and reinforcement learning. Each type serves distinct purposes and is suitable for different kinds of tasks.

1. Supervised Learning

Definition: In supervised learning, the model is trained on labeled data, where the input data is paired with the correct output. The objective is to learn a mapping from inputs to outputs.

Key Characteristics:

- **Training Data**: Requires a substantial amount of labeled training data.
- **Prediction**: The trained model can predict outcomes for new, unseen data.

Applications:

- **Classification**: Identifying which category an input belongs to (e.g., spam detection in emails).
- **Regression**: Predicting a continuous value (e.g., predicting house prices based on features like size and location).

Example: A common example of supervised learning is the use of decision trees to classify emails as either "spam" or "not spam" based on features extracted from the email content.

2. Unsupervised Learning

Definition: In unsupervised learning, the model is trained on data without labeled outputs. The goal is to identify patterns or groupings in the data.

Key Characteristics:

- **No Labels**: The training data does not have predefined labels or outcomes.
- **Pattern Recognition**: Focuses on discovering hidden structures within the data.

Applications:

- **Clustering**: Grouping similar data points together (e.g., customer segmentation in marketing).
- **Dimensionality Reduction**: Reducing the number of features while preserving essential information (e.g., Principal Component Analysis).

Example: An example of unsupervised learning is using k-means clustering to segment customers based on purchasing behavior, helping businesses tailor their marketing strategies.

3. Reinforcement Learning

Definition: Reinforcement learning (RL) is a type of machine learning where an agent learns to make decisions by taking actions in an environment to maximize cumulative rewards.

Key Characteristics:

- **Trial and Error**: The agent learns through interactions with the environment, receiving feedback in the form of rewards or penalties.
- **Long-Term Reward**: The focus is on maximizing long-term rewards rather than immediate outcomes.

Applications:

- **Game Playing**: Training agents to play games like chess or Go.
- **Robotics**: Enabling robots to learn complex tasks, such as navigation or manipulation.

Example: A well-known example of reinforcement learning is DeepMind's AlphaGo, which learned to play the game of Go at a superhuman level by playing millions of games against itself and refining its strategies.

Data Preparation and Feature Engineering

Before training a machine learning model, it is crucial to prepare the data adequately. This process includes data cleaning, transformation, and feature engineering.

1. Data Cleaning

Definition: Data cleaning involves identifying and correcting errors or inconsistencies in the dataset. This step ensures the quality and reliability of the data.

Common Techniques:

- **Handling Missing Values**: Filling in missing values or removing incomplete records.
- **Removing Duplicates**: Identifying and eliminating duplicate entries in the dataset.

2. Data Transformation

Definition: Data transformation includes modifying the dataset to improve the performance of machine learning models. This can involve scaling, normalization, and encoding categorical variables.

Common Techniques:

- **Normalization**: Scaling numeric features to a specific range (e.g., [0, 1]) to ensure that no feature dominates the learning process.
- **One-Hot Encoding**: Converting categorical variables into a numerical format suitable for ML algorithms.

3. Feature Engineering

Definition: Feature engineering is the process of creating new input features from existing data to improve model performance. This step is often crucial for the success of machine learning models.

Common Techniques:

- **Polynomial Features**: Creating new features by combining existing ones (e.g., x^2, xy).
- **Domain-Specific Features**: Extracting features based on domain knowledge (e.g., converting timestamps into day of the week or hour of the day).

Algorithms Used in AI Agents

Various algorithms are employed in AI agents to learn from data and make decisions. Below are some common algorithms categorized by their learning type.

1. Supervised Learning Algorithms

- **Linear Regression**: Used for predicting continuous outcomes based on linear relationships.
- **Logistic Regression**: Employed for binary classification tasks, predicting probabilities of classes.
- **Support Vector Machines (SVM)**: Effective for both classification and regression tasks, particularly in high-dimensional spaces.
- **Decision Trees**: A tree-like model used for classification and regression, which splits data based on feature values.

2. Unsupervised Learning Algorithms

- **K-Means Clustering**: Groups data points into k distinct clusters based on feature similarity.
- **Hierarchical Clustering**: Builds a hierarchy of clusters, allowing for multi-level analysis of data relationships.
- **Principal Component Analysis (PCA)**: A dimensionality reduction technique that transforms data into a lower-dimensional space.

3. Reinforcement Learning Algorithms

- **Q-Learning**: A model-free algorithm that learns the value of actions in particular states, enabling the agent to make informed decisions.
- **Deep Q-Networks (DQN)**: Combines Q-learning with deep neural networks to handle high-dimensional state spaces.
- **Proximal Policy Optimization (PPO)**: An advanced policy gradient method that balances exploration and exploitation for efficient learning.

Conclusion

Understanding machine learning is fundamental for mastering AI agents. By differentiating between supervised, unsupervised, and reinforcement learning, and recognizing the importance of data preparation and feature engineering, we lay the groundwork for developing effective AI agents. Furthermore, familiarity with the algorithms used in AI agents empowers developers and researchers to select the right approach for specific applications. As we continue through this book, we will explore natural language processing, computer vision, and other essential technologies that enhance AI agents, enabling them to perform increasingly complex tasks in diverse environments.

Chapter 6: Natural Language Processing for AI Agents

Natural Language Processing (NLP) is a crucial field within artificial intelligence that enables machines to understand, interpret, and generate human language in a way that is both meaningful and useful. As AI agents increasingly interact with users through spoken or written language, mastering NLP techniques becomes essential for enhancing their capabilities. This chapter provides an overview of key NLP techniques, explores sentiment analysis and language generation, and discusses various applications of NLP in conversational AI.

Overview of NLP Techniques

NLP encompasses a variety of techniques that facilitate the interaction between humans and machines through language. The following are some foundational techniques commonly used in NLP:

1. Tokenization

Definition: Tokenization is the process of breaking down text into smaller units called tokens, which can be words, phrases, or symbols. This is often the first step in NLP tasks.

Purpose: Tokenization helps in preparing the text for further analysis, allowing algorithms to process language effectively.

Example: In the sentence "I love AI agents," tokenization would break it down into the tokens: ["I", "love", "AI", "agents"].

2. Part-of-Speech Tagging

Definition: Part-of-speech (POS) tagging assigns a grammatical category (noun, verb, adjective, etc.) to each token in a sentence.

Purpose: Understanding the grammatical structure of sentences enables agents to better interpret meaning and context.

Example: For the sentence "The cat sat on the mat," the POS tagging would identify "The" (determiner), "cat" (noun), "sat" (verb), "on" (preposition), "the" (determiner), and "mat" (noun).

3. Named Entity Recognition (NER)

Definition: NER involves identifying and classifying key entities in text into predefined categories such as names of people, organizations, locations, dates, and more.

Purpose: This technique is essential for extracting important information from text and understanding context.

Example: In the sentence "Apple Inc. was founded in Cupertino, California," NER would recognize "Apple Inc." as an organization and "Cupertino" and "California" as locations.

4. Sentiment Analysis

Definition: Sentiment analysis determines the emotional tone behind a series of words, identifying whether the sentiment is positive, negative, or neutral.

Purpose: This technique is useful for understanding opinions and attitudes expressed in text, which is particularly valuable in customer feedback and social media monitoring.

Example: The phrase "I love this product!" would be classified as positive sentiment, while "I hate waiting for customer service" would be classified as negative sentiment.

5. Language Generation

Definition: Language generation involves creating meaningful text based on specific input or context. This can range from simple responses to complex narrative generation.

Purpose: This technique allows AI agents to produce human-like text, enabling more engaging interactions.

Example: A chatbot might generate a response like "Thank you for your feedback! We appreciate your support," based on user input.

Understanding Sentiment Analysis and Language Generation

1. Sentiment Analysis in Depth

Sentiment analysis typically involves several steps:

- **Data Collection**: Gathering text data from various sources, such as social media posts, product reviews, or customer feedback.
- **Preprocessing**: Cleaning the data by removing irrelevant information, normalizing text (e.g., converting to lowercase), and tokenizing the input.
- **Feature Extraction**: Transforming text into numerical features that can be fed into machine learning models. This often involves techniques like bag-of-words, term frequency-inverse document frequency (TF-IDF), or word embeddings (e.g., Word2Vec, GloVe).
- **Model Training**: Using labeled datasets to train machine learning models to classify sentiment.
- **Evaluation**: Testing the model's accuracy and refining it based on performance metrics such as precision, recall, and F1-score.

2. Language Generation Techniques

Language generation can be achieved through various methods:

- **Template-Based Generation**: Involves using predefined templates to fill in specific variables. While straightforward, it can lead to repetitive or unnatural language.
- **Statistical Language Models**: These models, such as n-grams, predict the next word based on the previous words. They have limitations in capturing long-term dependencies in language.
- **Neural Language Models**: More advanced approaches, like recurrent neural networks (RNNs) and transformer models (e.g., GPT-3), leverage deep learning to generate coherent and contextually relevant text.

Applications in Conversational AI

Natural Language Processing is integral to the development of conversational AI systems. Below are some key applications:

1. Chatbots

Role: Chatbots utilize NLP techniques to understand user queries and provide relevant responses. They can be deployed in customer support, e-commerce, and information retrieval.

Example: A travel booking chatbot that assists users in finding flights and hotels based on their preferences, utilizing sentiment analysis to gauge customer satisfaction.

2. Virtual Assistants

Role: Virtual assistants like Siri, Alexa, and Google Assistant leverage NLP to interpret voice commands and perform tasks such as setting reminders, playing music, or providing weather updates.

Example: A user might say, "What's the weather like today?" and the assistant responds with the current conditions and forecast.

3. Social Media Monitoring

Role: Businesses employ sentiment analysis tools to monitor social media platforms for brand mentions, assessing public opinion and feedback.

Example: Analyzing tweets about a product launch to gauge customer sentiment and adjust marketing strategies accordingly.

Conclusion

Natural Language Processing is a fundamental aspect of developing intelligent AI agents capable of understanding and generating human language. By mastering key NLP techniques such as sentiment analysis and language generation, developers can create more engaging and effective conversational agents. As we progress in this book, we will explore other critical technologies like computer vision and multi-agent systems, further expanding our understanding of AI agents and their capabilities in various domains.

Chapter 7: Computer Vision and AI Agents

Computer vision is an essential field of artificial intelligence that enables machines to interpret and understand visual information from the world. By leveraging computer vision techniques, AI agents can analyze images, recognize objects, and even make decisions based on visual data. This chapter provides an overview of the basics of computer vision, discusses image recognition and processing techniques, and explores use cases in robotics and autonomous vehicles.

Basics of Computer Vision

Definition

Computer vision involves the extraction, analysis, and interpretation of information from images or video. It combines elements from various disciplines, including computer science, mathematics, and cognitive science, to enable machines to process visual data similarly to how humans do.

Key Components

1. **Image Acquisition**: The process of capturing images through cameras or sensors. This can involve still images, video streams, or 3D data.
2. **Image Processing**: The manipulation of images to enhance quality or extract useful features. Techniques include filtering, resizing, and adjusting brightness and contrast.
3. **Feature Extraction**: Identifying and extracting important features from images, such as edges, shapes, or textures, which are crucial for further analysis.
4. **Pattern Recognition**: The identification of patterns or objects in images based on the features extracted. This can involve classification, detection, or segmentation tasks.

Image Recognition and Processing
Image Recognition

Definition: Image recognition is a subset of computer vision focused on identifying and classifying objects within images. It allows AI agents to understand the content of visual inputs.

Techniques:

- **Convolutional Neural Networks (CNNs)**: A deep learning architecture designed for processing grid-like data such as images. CNNs excel in automatically detecting features through multiple layers of convolutional and pooling operations.
- **Transfer Learning**: A technique that utilizes pre-trained models on large datasets to improve performance on smaller, specific tasks. This approach is beneficial when labeled data is scarce.

Applications:

- **Facial Recognition**: Identifying or verifying individuals in images by analyzing facial features.
- **Object Detection**: Locating and classifying multiple objects within an image, as seen in applications like autonomous driving.

Image Processing

Definition: Image processing refers to techniques used to enhance or manipulate images to improve their quality or extract relevant information.

Common Techniques:

- **Filtering**: Removing noise or enhancing specific features using filters (e.g., Gaussian, median).
- **Edge Detection**: Identifying the boundaries of objects within images, commonly using algorithms like the Canny edge detector.
- **Image Segmentation**: Dividing an image into meaningful segments to simplify analysis, such as separating an object from its background.

Use Cases in Robotics and Autonomous Vehicles

1. Robotics

Robots equipped with computer vision can perceive and interact with their environments effectively. Key applications include:

- **Navigation**: Robots use visual inputs to navigate through complex environments, avoiding obstacles and following predefined paths.
- **Object Manipulation**: Visual recognition allows robots to identify and grasp objects, facilitating tasks like sorting, assembly, or delivery.

Example: Robotic arms in manufacturing use computer vision to locate and pick up items on a conveyor belt, improving efficiency and accuracy in production lines.

2. Autonomous Vehicles

Computer vision is a foundational technology for self-driving cars, enabling them to interpret their surroundings. Key applications include:

- **Object Detection and Classification**: Identifying pedestrians, vehicles, traffic signs, and road conditions in real-time to make informed driving decisions.
- **Lane Detection**: Analyzing road markings to maintain lane discipline and prevent accidents.
- **Navigation and Mapping**: Utilizing visual data to create maps of the environment, assisting in route planning and obstacle avoidance.

Example: Companies like Waymo and Tesla leverage advanced computer vision algorithms in their autonomous vehicles to ensure safe navigation through urban environments.

Conclusion

Computer vision plays a critical role in enhancing the capabilities of AI agents, enabling them to process and interpret visual information effectively. By understanding the fundamentals of computer vision, including image recognition and processing techniques, developers can create more sophisticated AI agents capable of operating in dynamic environments. As we move forward in this book, we will explore multi-agent systems and reinforcement learning, further expanding our understanding of how AI agents can collaborate and learn from their experiences in complex scenarios. The integration of computer vision with AI agents holds immense potential for innovative applications across various industries, paving the way for a more intelligent future.

Chapter 8: Multi-Agent Systems

Multi-agent systems (MAS) represent a complex yet fascinating area within the field of artificial intelligence, where multiple agents interact, collaborate, and sometimes compete to achieve their individual or collective goals. This chapter introduces the concept of multi-agent systems, explores the communication and collaboration mechanisms between agents, and examines their applications across various industries.

Introduction to Multi-Agent Systems (MAS)

Definition

A multi-agent system is defined as a system composed of multiple interacting intelligent agents that can work independently or collaboratively to achieve tasks or solve problems. These agents can be autonomous, meaning they operate without direct human intervention, and they may possess varying degrees of intelligence and capabilities.

Characteristics of Multi-Agent Systems

1. **Autonomy**: Each agent in a MAS operates independently, making its own decisions based on its perception of the environment.
2. **Cooperation**: Agents may need to collaborate to achieve common goals, sharing information and resources.
3. **Communication**: Agents communicate with each other using predefined protocols or languages, facilitating coordination and information exchange.
4. **Adaptability**: Agents can adapt to changes in the environment or in the behavior of other agents, enhancing the overall system's resilience and efficiency.
5. **Distributed Nature**: The agents in a MAS are often distributed across different locations, allowing for scalability and flexibility.

Communication and Collaboration Between Agents
Communication Mechanisms

Effective communication is crucial for agents to collaborate efficiently. Various mechanisms and protocols facilitate communication in multi-agent systems:

1. **Message Passing**: Agents send and receive messages to exchange information, requests, or commands. This can be synchronous (real-time) or asynchronous.
2. **Shared Knowledge Bases**: Agents can access a common knowledge base, allowing them to retrieve and update information collaboratively.
3. **Communication Languages**: Standardized languages, such as Agent Communication Language (ACL) or FIPA-ACL, provide structured formats for agents to communicate their intentions, actions, and observations.

Collaboration Strategies

Collaboration among agents can take many forms, depending on the nature of the tasks and the goals of the system:

1. **Task Distribution**: Agents can divide tasks among themselves based on their capabilities, ensuring that work is done efficiently and effectively.
2. **Negotiation**: Agents may negotiate with each other to reach agreements on resource allocation, task assignments, or conflict resolution.
3. **Coordination**: Agents coordinate their actions to achieve a common goal while minimizing conflicts or redundancies. This can involve techniques like shared plans or joint intentions.
4. **Competitive Interactions**: In some systems, agents may compete for resources or tasks, which can lead to innovative solutions and improvements in performance.

Applications of Multi-Agent Systems in Various Industries

Multi-agent systems are utilized in a wide range of applications across different sectors. Below are some prominent examples:

1. Robotics

In robotics, multi-agent systems enable teams of robots to work together to accomplish complex tasks, such as search and rescue missions, exploration, and surveillance.

Example: Swarm robotics involves multiple simple robots that collaborate to perform tasks like environmental monitoring or agricultural applications, mimicking the behavior of social insects.

2. Smart Transportation

Multi-agent systems can optimize traffic management by coordinating vehicles and traffic signals to reduce congestion and improve safety.

Example: Autonomous vehicles can communicate with each other and with traffic infrastructure, sharing information about traffic conditions, road hazards, and optimal routes.

3. Smart Grids

In energy management, multi-agent systems facilitate the coordination of distributed energy resources, such as solar panels and battery storage systems, to optimize energy production and consumption.

Example: Agents representing different energy producers and consumers can negotiate energy trades, manage load balancing, and respond to changes in demand.

4. Healthcare

Multi-agent systems can enhance healthcare delivery by coordinating patient care, resource allocation, and data sharing among healthcare providers.

Example: Agents in a hospital can manage patient scheduling, track medication administration, and facilitate communication between doctors and nurses to improve patient outcomes.

5. E-commerce

In e-commerce, multi-agent systems can optimize supply chain management by allowing agents to represent suppliers, manufacturers, and retailers, facilitating negotiations and inventory management.

Example: Agents can dynamically adjust pricing based on demand, manage stock levels, and negotiate contracts, improving overall efficiency in the supply chain.

Conclusion

Multi-agent systems offer powerful frameworks for tackling complex problems that require collaboration and communication between multiple intelligent agents. By understanding the principles of MAS, including communication mechanisms and collaboration strategies, we can unlock their potential across various industries. As we advance in this book, we will explore reinforcement learning and its applications, delving deeper into how AI agents learn and adapt in dynamic environments. The future of AI agents and multi-agent systems is bright, promising innovative solutions and enhanced capabilities across diverse domains.

Chapter 9: Reinforcement Learning

Reinforcement Learning (RL) is a dynamic and fascinating subfield of machine learning that empowers AI agents to learn from their interactions with an environment in order to achieve specific goals. Unlike traditional supervised learning, where models are trained on labeled datasets, RL focuses on learning optimal behaviors through trial and error, guided by feedback in the form of rewards or penalties. This chapter delves into the principles of reinforcement learning, the concept of Markov Decision Processes (MDP), and presents several compelling case studies that illustrate the effectiveness of RL in training AI agents.

Principles of Reinforcement Learning

1. The Learning Process

Reinforcement learning involves an agent that interacts with an environment, makes decisions, and receives feedback. The primary components of the RL framework include:

- **Agent**: The learner or decision maker.
- **Environment**: The world with which the agent interacts.
- **Actions**: The set of all possible moves the agent can make in the environment.
- **States**: The different configurations of the environment as perceived by the agent.
- **Rewards**: Feedback from the environment based on the actions taken by the agent, which can be positive or negative.

2. Exploration vs. Exploitation

A central challenge in reinforcement learning is balancing exploration and exploitation:

- **Exploration**: The agent tries new actions to discover their effects, which may lead to obtaining better rewards.
- **Exploitation**: The agent utilizes known actions that have previously yielded high rewards, optimizing its performance based on existing knowledge.

Finding the right balance between these two strategies is crucial for effective learning.

3. Policy

A policy defines the behavior of an agent, mapping states of the environment to actions. There are two types of policies:

- **Deterministic Policy**: The agent takes a specific action for a given state.
- **Stochastic Policy**: The agent selects actions based on a probability distribution over possible actions for each state.

The objective of reinforcement learning is to discover an optimal policy that maximizes cumulative rewards over time.

4. Value Function

The value function estimates how good it is for an agent to be in a given state. It helps the agent understand the long-term reward potential of states, which aids in decision-making. There are two main types of value functions:

- **State Value Function (V)**: The expected return for being in a particular state and following a policy thereafter.
- **Action Value Function (Q)**: The expected return for taking a specific action in a given state and then following a policy.

Markov Decision Processes (MDP)

1. Definition of MDP

A Markov Decision Process (MDP) provides a mathematical framework for modeling decision-making in situations where outcomes are partly random and partly under the control of a decision maker. MDPs consist of the following components:

- **States (S)**: A finite set of states that represent all possible situations in the environment.
- **Actions (A)**: A finite set of actions available to the agent in each state.
- **Transition Probabilities (P)**: The probability of transitioning from one state to another given an action.
- **Rewards (R)**: The immediate reward received after transitioning to a new state due to an action.
- **Discount Factor (γ)**: A factor between 0 and 1 that determines the importance of future rewards compared to immediate rewards.

2. Solving MDPs

The goal of solving an MDP is to find an optimal policy that maximizes the expected cumulative reward. Common algorithms for solving MDPs include:

- **Value Iteration**: An iterative method that updates the value of each state until convergence, from which an optimal policy can be derived.
- **Policy Iteration**: Alternates between policy evaluation (calculating the value of a given policy) and policy improvement (updating the policy based on current value estimates).

Case Studies of Reinforcement Learning in AI Agents

1. AlphaGo

One of the most celebrated applications of reinforcement learning is AlphaGo, an AI program developed by DeepMind to play the board game Go. By using a combination of deep neural networks and reinforcement learning techniques, AlphaGo achieved superhuman performance:

- **Approach**: AlphaGo used a combination of supervised learning from human games and reinforcement learning from self-play, continuously improving its strategies.
- **Outcome**: AlphaGo defeated world champion Go player Lee Sedol in 2016, showcasing the power of RL in mastering complex games.

2. Autonomous Driving

Reinforcement learning has also made significant strides in the realm of autonomous vehicles:

- **Approach**: Companies like Waymo and Tesla employ RL techniques to train self-driving cars, allowing them to learn optimal driving policies through simulated environments.
- **Outcome**: These AI agents adapt to various driving scenarios, improving safety and efficiency on the road by continuously learning from their interactions.

3. Robotics

In robotics, reinforcement learning enables robots to perform complex tasks:

- **Approach**: Robots can learn to navigate obstacles, pick and place objects, or manipulate items through trial and error in a controlled environment.
- **Outcome**: These robots enhance their capabilities over time, adapting to new tasks and environments without needing extensive reprogramming.

Conclusion

Reinforcement learning is a powerful paradigm that empowers AI agents to learn and adapt through experience, driving innovations across various fields. By understanding the principles of RL, including the critical role of Markov Decision Processes and practical applications in real-world scenarios, developers and researchers can harness the full potential of AI agents. As we proceed through the book, we will explore ethical considerations in AI, as well as the implications of these technologies in different domains, ensuring that we approach AI development responsibly and thoughtfully.

Chapter 10: Ethical Considerations in AI

As artificial intelligence continues to evolve and permeate various aspects of society, ethical considerations surrounding its development and deployment become increasingly critical. The integration of AI agents into everyday life raises numerous ethical dilemmas that must be addressed to ensure fairness, accountability, and transparency. This chapter explores the understanding of bias and fairness in AI, ethical dilemmas in AI decision-making, and frameworks for responsible AI development.

Understanding Bias and Fairness in AI

1. Bias in AI

Bias in AI refers to systematic errors in algorithms that produce unfair outcomes, disproportionately affecting certain groups of people. This bias can arise from various sources:

- **Data Bias**: AI systems are trained on data that may reflect historical inequalities or societal biases. If the training data is not representative, the resulting AI agent may perpetuate or even exacerbate these biases.
- **Algorithmic Bias**: The algorithms used in AI systems can also introduce bias. For example, certain design choices or optimization criteria may favor specific outcomes over others, leading to unfair treatment of particular groups.
- **Human Bias**: Biases can be inadvertently introduced by developers during the design and implementation stages, influenced by their own beliefs and experiences.

2. Fairness in AI

Fairness in AI aims to ensure that outcomes from AI systems are equitable and just. There are several definitions of fairness, which can lead to different implications for AI development:

- **Individual Fairness**: Similar individuals should receive similar outcomes. This approach emphasizes that fairness is based on equal treatment of individuals.
- **Group Fairness**: Outcomes should be equal across predefined demographic groups (e.g., gender, race). This approach seeks to ensure that no group is disproportionately favored or disadvantaged.
- **Subgroup Fairness**: Similar to group fairness, but focuses on subgroups within larger populations, ensuring that certain marginalized communities receive equitable treatment.

Ethical Dilemmas in AI Decision-Making

1. Accountability

Determining accountability for decisions made by AI agents poses significant ethical challenges. If an AI system makes a mistake—such as misidentifying an individual or making an erroneous recommendation—who is responsible? Is it the developer, the organization deploying the AI, or the AI itself? This ambiguity complicates issues of liability and recourse for affected individuals.

2. Transparency

Transparency in AI systems is vital for fostering trust and understanding among users. However, many AI algorithms, particularly those based on deep learning, function as "black boxes," making it challenging to understand how decisions are made. This lack of transparency raises ethical concerns, as users may not be able to assess the fairness or reliability of AI-driven outcomes.

3. Informed Consent

When deploying AI agents, especially in sensitive areas such as healthcare, users must be adequately informed about how their data will be used and the implications of AI-assisted decisions. Ethical concerns arise if individuals are not given a clear understanding of the risks, benefits, and potential biases associated with AI technologies.

4. Autonomy and Manipulation

AI agents often influence user behavior, particularly in applications such as recommendation systems and advertising. Ethical dilemmas arise when AI systems manipulate users into making decisions that may not be in their best interests. Striking a balance between personalized experiences and respecting user autonomy is critical.

Frameworks for Responsible AI Development

To navigate the ethical landscape of AI, several frameworks and guidelines have been proposed to guide developers and organizations in creating responsible AI systems:

1. Ethical Guidelines and Principles

Organizations such as the IEEE and the European Commission have developed ethical guidelines outlining principles for AI development. Key principles include:

- **Fairness**: Ensure that AI systems treat all users equitably and do not reinforce existing biases.
- **Accountability**: Establish clear lines of responsibility for AI decision-making.
- **Transparency**: Make the workings of AI systems comprehensible and accessible to users.
- **Privacy**: Protect user data and ensure that it is used responsibly.

2. Ethical Audits

Conducting regular ethical audits of AI systems can help identify potential biases and ethical concerns. These audits assess the data, algorithms, and outcomes of AI systems to ensure compliance with ethical standards and principles.

3. Interdisciplinary Collaboration

Addressing ethical considerations in AI requires collaboration across various disciplines, including computer science, ethics, law, and social sciences. Involving diverse perspectives in the development process can help identify and mitigate ethical risks.

4. User Engagement and Feedback

Engaging users throughout the AI development process and incorporating their feedback can enhance transparency and accountability. By understanding user perspectives and concerns, developers can create AI agents that align more closely with ethical standards and user expectations.

Conclusion

As AI agents become more integrated into our daily lives, the ethical considerations surrounding their development and deployment cannot be overlooked. Understanding bias and fairness, addressing ethical dilemmas in decision-making, and implementing frameworks for responsible AI development are crucial steps in ensuring that AI technologies are developed and used in a manner that promotes equity and justice. In the following chapters, we will explore the role of AI in automation and various industries, emphasizing the importance of ethical considerations in these applications. By fostering a responsible approach to AI development, we can harness the potential of AI agents to improve our lives while safeguarding fundamental ethical principles.

Chapter 11: The Role of AI in Automation

As businesses and industries seek to enhance efficiency and productivity, artificial intelligence (AI) agents have emerged as pivotal tools in the automation landscape. This chapter delves into the concepts of automation and augmentation, exploring how AI agents drive productivity, as well as providing case studies that highlight their impact across various sectors.

1. Automation vs. Augmentation

1.1 Understanding Automation

Automation refers to the technology-driven process that enables machines, including AI agents, to perform tasks without human intervention. This can involve automating repetitive tasks, streamlining workflows, and enhancing decision-making processes. The primary objective of automation is to increase efficiency, reduce errors, and minimize labor costs.

1.2 The Concept of Augmentation

While automation focuses on replacing human labor, augmentation emphasizes enhancing human capabilities with AI. In this model, AI agents assist humans in tasks that require judgment, creativity, or emotional intelligence. Augmentation seeks to create a synergistic relationship between humans and machines, leveraging the strengths of both.

1.3 The Balance of Automation and Augmentation

In many industries, a combination of automation and augmentation is ideal. By automating routine tasks, human workers can focus on higher-level responsibilities, improving overall productivity. For instance, in customer service, AI chatbots can handle inquiries, allowing human agents to focus on complex issues that require empathy and critical thinking.

2. How AI Agents Enhance Productivity

AI agents contribute to productivity in various ways:

2.1 Streamlining Processes

AI agents can analyze vast amounts of data, identify patterns, and optimize workflows. For example, in manufacturing, AI-driven robots can monitor production lines, detect inefficiencies, and adjust operations in real time to enhance productivity.

2.2 Reducing Errors

Human error is a significant factor in many industries. AI agents, programmed to follow precise algorithms, minimize the likelihood of mistakes. In finance, for instance, AI can process transactions and analyze data with a level of accuracy that far exceeds human capabilities.

2.3 Enabling Scalability

AI agents allow businesses to scale their operations quickly. By automating processes, organizations can handle increased workloads without the need for proportional increases in staff. This flexibility is especially valuable in industries such as e-commerce, where demand can fluctuate rapidly.

2.4 Enhancing Decision-Making

AI agents can provide insights based on data analysis, improving decision-making. In healthcare, AI systems can analyze patient data to assist doctors in diagnosing conditions and recommending treatments, leading to better patient outcomes.

3. Case Studies in Industry Automation

3.1 Manufacturing

In the automotive industry, companies like Ford and General Motors have adopted AI-powered robots to automate assembly lines. These robots perform tasks such as welding, painting, and assembly with precision and speed. The integration of AI has resulted in significant cost savings and increased production rates, allowing manufacturers to meet consumer demand more efficiently.

3.2 Retail

AI agents are transforming retail by enhancing inventory management and customer experience. Amazon's use of AI-driven robots in warehouses optimizes storage and order fulfillment, enabling quicker delivery times. Additionally, AI chatbots on e-commerce platforms provide instant customer support, helping customers find products and answer inquiries without human intervention.

3.3 Healthcare

In healthcare, AI agents are streamlining administrative tasks and improving patient care. Hospitals utilize AI for scheduling appointments, managing patient records, and billing processes, reducing administrative burdens on staff. AI algorithms analyze medical images to assist radiologists in detecting anomalies, thus enhancing diagnostic accuracy.

3.4 Finance

AI agents have revolutionized the finance sector by automating trading processes and enhancing risk assessment. Companies like BlackRock use AI to analyze market trends and make investment decisions at a speed and accuracy unattainable by human traders. Furthermore, AI-driven fraud detection systems analyze transactions in real-time, identifying suspicious activity and protecting consumers.

4. Challenges and Considerations

While AI agents offer significant benefits in automation, there are challenges to consider:

4.1 Job Displacement

The rise of automation raises concerns about job displacement. As AI agents take over repetitive tasks, workers may find themselves out of work. Organizations must focus on retraining and upskilling employees to adapt to new roles that AI technologies create.

4.2 Dependence on Technology

Increased reliance on AI agents can lead to vulnerabilities. Organizations must ensure robust cybersecurity measures are in place to protect sensitive data and maintain operational integrity.

4.3 Ethical Considerations

As automation expands, ethical considerations regarding decision-making processes must be addressed. Organizations must ensure that AI systems are designed to operate fairly and transparently, minimizing biases in automated decisions.

Conclusion

AI agents are reshaping the landscape of automation by enhancing productivity across various industries. The balance between automation and augmentation is crucial in maximizing the benefits of AI while addressing potential challenges. As organizations continue to integrate AI into their operations, understanding the role of AI agents will be essential for navigating the evolving workforce and ensuring responsible implementation. The subsequent chapters will delve deeper into specific applications of AI agents across different sectors, further illustrating their transformative potential.

Chapter 12: AI Agents in Business

In today's rapidly evolving business landscape, AI agents have become crucial in driving efficiency, enhancing customer experiences, and enabling data-driven decision-making. This chapter explores the diverse applications of AI agents in marketing, sales, customer service, and other business domains, focusing on personalization, recommendation systems, and measuring the return on investment (ROI) of AI implementations.

1. Applications of AI Agents in Business
1.1 Marketing

AI agents play a transformative role in marketing strategies. By analyzing vast amounts of data, they provide insights into consumer behavior, preferences, and trends, allowing businesses to tailor their marketing efforts effectively.

- **Customer Segmentation:** AI agents can segment customers based on their behavior, demographics, and purchasing history, enabling businesses to target specific groups with personalized marketing campaigns.
- **Predictive Analytics:** AI algorithms can predict future buying behaviors and trends, helping marketers to craft timely campaigns that resonate with their audience.
- **Content Generation:** AI tools can generate marketing content, from social media posts to email campaigns, optimizing them for engagement and conversion.

1.2 Sales

AI agents enhance the sales process by streamlining workflows and improving customer interactions. They assist sales teams in various ways:

- **Lead Scoring:** AI agents can analyze leads based on various factors, scoring them to prioritize those with the highest conversion potential. This enables sales teams to focus their efforts on the most promising opportunities.
- **Sales Forecasting:** By analyzing historical sales data and market conditions, AI can provide accurate sales forecasts, helping businesses plan their inventory and resources effectively.
- **Virtual Sales Assistants:** AI-powered chatbots can engage with potential customers on websites, answer queries, and guide them through the sales funnel, significantly enhancing the customer experience.

1.3 Customer Service

AI agents have revolutionized customer service by providing efficient and effective solutions:

- **Chatbots and Virtual Assistants:** These AI agents can handle a multitude of customer inquiries simultaneously, providing instant responses and freeing human agents to focus on complex issues. They can assist with frequently asked questions, order tracking, and troubleshooting.
- **Sentiment Analysis:** AI can analyze customer interactions to assess sentiment, enabling businesses to address issues proactively and improve customer satisfaction.
- **24/7 Availability:** AI agents can operate around the clock, ensuring that customers receive support whenever they need it, thus enhancing the overall customer experience.

2. Personalization and Recommendation Systems
2.1 Importance of Personalization

Personalization is critical in today's competitive business environment. AI agents analyze user data to deliver tailored experiences that enhance customer satisfaction and loyalty.

- **Personalized Recommendations:** AI algorithms power recommendation engines that suggest products or services based on a customer's browsing history, preferences, and past purchases. Companies like Amazon and Netflix excel in this area, leveraging AI to keep users engaged and drive sales.
- **Dynamic Content:** AI can personalize content on websites and emails, adapting offers and messages to align with individual user profiles, resulting in higher engagement rates.

2.2 Case Study: E-Commerce Personalization

In the e-commerce sector, AI agents have significantly improved conversion rates through personalized shopping experiences. By utilizing machine learning algorithms, companies analyze user behavior to offer tailored recommendations. For instance, an online clothing retailer might use AI to suggest items based on a user's previous purchases, resulting in increased sales and customer satisfaction.

3. Measuring ROI of AI Agent Implementations
3.1 Understanding ROI

Measuring the return on investment (ROI) for AI implementations is essential for justifying expenditures and demonstrating value to stakeholders. Key metrics to consider include:

- **Cost Savings:** Assess the reduction in operational costs achieved through automation and efficiency gains from AI agents.
- **Increased Revenue:** Evaluate the uplift in sales attributed to AI-driven marketing, sales, and customer service strategies.
- **Customer Satisfaction:** Analyze improvements in customer satisfaction scores, Net Promoter Scores (NPS), and customer retention rates.

3.2 Strategies for Measuring ROI

To effectively measure ROI, businesses should:

- **Set Clear Objectives:** Define specific, measurable goals for AI implementations, such as sales targets or customer service response times.
- **Track Key Performance Indicators (KPIs):** Monitor relevant KPIs throughout the implementation process to assess performance against established objectives.
- **Conduct Comparative Analysis:** Compare pre-implementation metrics with post-implementation performance to quantify the impact of AI agents.

4. Conclusion

AI agents are revolutionizing business operations across various sectors. From enhancing marketing strategies to improving customer service, their ability to analyze data and automate processes drives significant value. As businesses continue to embrace AI technology, understanding the applications, benefits, and ROI of AI agents will be crucial for maintaining a competitive edge in the evolving marketplace. The subsequent chapters will further explore the role of AI agents in specific industries, providing insights into their diverse applications and transformative potential.

Chapter 13: AI Agents in Healthcare

AI agents are transforming the healthcare landscape, offering innovative solutions that improve patient care, streamline operations, and enhance decision-making processes. This chapter explores the various applications of AI agents in healthcare, focusing on diagnostics, treatment, patient monitoring, virtual health assistants, and the ethical considerations that accompany these advancements.

1. Use Cases of AI Agents in Diagnostics and Treatment

1.1 Diagnostic Assistance

AI agents are increasingly employed in diagnostic processes, providing healthcare professionals with tools that enhance accuracy and efficiency. For example:

- **Medical Imaging Analysis:** AI algorithms analyze medical images—such as X-rays, MRIs, and CT scans—to identify abnormalities. Techniques like deep learning enable these agents to detect conditions such as tumors, fractures, and other anomalies with high precision. Research has shown that AI can outperform human radiologists in certain diagnostic tasks, significantly reducing the time needed for analysis.
- **Pathology:** AI agents assist pathologists by analyzing tissue samples and identifying cancerous cells. This not only speeds up the diagnostic process but also reduces human error, ensuring that more patients receive timely and accurate diagnoses.

1.2 Personalized Treatment Plans

AI agents help in tailoring treatment plans based on individual patient data:

- **Predictive Analytics:** By analyzing historical patient data, AI agents can predict how different patients will respond to specific treatments. This allows for the development of personalized treatment strategies that improve patient outcomes and reduce adverse effects.
- **Drug Discovery:** AI technologies streamline the drug discovery process by predicting how various compounds will interact with biological systems. AI agents can identify potential drug candidates faster than traditional methods, expediting the journey from laboratory research to clinical trials.

2. Patient Monitoring and Virtual Health Assistants

2.1 Remote Patient Monitoring

AI agents facilitate continuous patient monitoring, allowing healthcare providers to track patient health metrics in real-time:

- **Wearable Devices:** AI-powered wearables collect and analyze data such as heart rate, blood pressure, and glucose levels. These agents can alert healthcare providers to significant changes in a patient's condition, enabling timely interventions.
- **Chronic Disease Management:** For patients with chronic conditions, AI agents can provide ongoing support and monitoring. They can send reminders for medication, alert patients to necessary lifestyle adjustments, and notify healthcare providers if a patient's condition deteriorates.

2.2 Virtual Health Assistants

Virtual health assistants, powered by AI, offer a range of functionalities:

- **Symptom Checkers:** AI agents can analyze patient-reported symptoms and provide recommendations for next steps, such as whether to seek medical attention. This helps triage patients and optimize healthcare resources.
- **Appointment Scheduling:** AI assistants can manage scheduling for healthcare providers, streamlining the administrative burden and improving patient access to care.
- **Patient Education:** AI agents can deliver personalized health information to patients, enhancing their understanding of conditions and treatments, and empowering them to take an active role in their healthcare.

3. Ethical Considerations in Healthcare AI

As AI agents continue to integrate into healthcare systems, several ethical considerations arise:

3.1 Bias and Fairness

AI systems can inherit biases from the data they are trained on. If the training data does not adequately represent diverse populations, AI agents may provide less accurate predictions for underrepresented groups, leading to disparities in healthcare outcomes.

3.2 Privacy and Security

Patient data is highly sensitive, and AI systems that handle this information must ensure strict privacy protections. Safeguarding against data breaches and ensuring compliance with regulations such as HIPAA (Health Insurance Portability and Accountability Act) are crucial for maintaining patient trust.

3.3 Accountability

As AI agents take on more decision-making roles, questions arise about accountability. Determining who is responsible for errors made by AI—be it the developers, healthcare providers, or the institutions using the technology—remains a complex issue.

3.4 Informed Consent

The use of AI in healthcare raises questions about informed consent. Patients should be made aware when AI systems are involved in their care and understand how their data will be used. Clear communication about the benefits and risks of AI-assisted treatments is essential.

4. Conclusion

AI agents are poised to significantly enhance the healthcare sector, from improving diagnostic accuracy to providing personalized treatment and patient support. However, as these technologies become more prevalent, it is essential to address the ethical implications and ensure that AI is implemented responsibly and equitably. The next chapters will explore specific applications of AI agents in finance, business, and other industries, showcasing their transformative potential across various domains.

Chapter 14: AI Agents in Finance

AI agents are revolutionizing the finance industry by providing tools and technologies that enhance efficiency, accuracy, and decision-making. From algorithmic trading to fraud detection, the integration of AI agents in finance allows for innovative solutions that can analyze vast amounts of data in real-time, predict market trends, and automate various processes. This chapter explores the critical applications of AI agents in finance, focusing on algorithmic trading, risk assessment, and regulatory considerations.

1. Algorithmic Trading and Financial Forecasting

1.1 Overview of Algorithmic Trading

Algorithmic trading involves using AI agents to execute trades based on predefined criteria, which can include price, volume, and timing. These agents analyze market data and execute orders at speeds and frequencies that are impossible for human traders.

- **High-Frequency Trading (HFT):** A subset of algorithmic trading that uses complex algorithms to analyze multiple markets and execute orders based on market conditions within fractions of a second. HFT can capitalize on small price discrepancies that may exist for only a moment.
- **Quantitative Trading:** This strategy utilizes mathematical models and algorithms to identify trading opportunities. AI agents can develop these models based on historical data, providing traders with insights and predictions that enhance decision-making.

1.2 Financial Forecasting

AI agents also play a crucial role in financial forecasting, using machine learning techniques to predict market movements and economic trends. By analyzing historical data and identifying patterns, these agents can provide valuable insights for investment strategies.

- **Time Series Analysis:** AI agents can use time series forecasting to analyze historical price data, taking into account seasonality, trends, and cycles to predict future movements in asset prices.
- **Sentiment Analysis:** AI can process news articles, social media, and other textual data to gauge market sentiment. By understanding public sentiment, AI agents can help predict price movements based on external factors affecting investor behavior.

2. Risk Assessment and Fraud Detection

2.1 Risk Assessment

AI agents are increasingly used in assessing credit and market risks. By analyzing a multitude of variables, including economic indicators and historical data, these agents can provide financial institutions with a clearer picture of potential risks.

- **Credit Scoring:** AI-driven credit scoring models can analyze a wider range of factors than traditional models, allowing for more accurate assessments of borrower creditworthiness. This reduces the risk of default and helps lenders make more informed decisions.
- **Portfolio Management:** AI agents assist in managing investment portfolios by assessing risk and optimizing asset allocation. They can analyze market conditions and suggest adjustments to maintain the desired risk level in a portfolio.

2.2 Fraud Detection

Fraud detection is a critical area where AI agents excel, providing advanced capabilities to identify suspicious activities and transactions.

- **Anomaly Detection:** AI agents can continuously monitor transactions in real-time, using machine learning algorithms to identify patterns that deviate from the norm. If an anomaly is detected—such as unusual spending behavior—the system can flag the transaction for further investigation.
- **Behavioral Analytics:** By establishing baseline behaviors for users, AI agents can detect when a user's actions fall outside their typical patterns. This is particularly effective for identifying potential identity theft or account takeover attempts.

3. Regulatory Considerations in Financial AI

The integration of AI agents in finance also brings about significant regulatory considerations. As these technologies evolve, regulatory bodies are tasked with ensuring that they are used responsibly and ethically.

3.1 Compliance with Regulations

Financial institutions must ensure that their AI systems comply with existing regulations, including anti-money laundering (AML) and Know Your Customer (KYC) requirements. AI agents can assist in maintaining compliance by automating data collection and reporting.

3.2 Transparency and Accountability

One of the challenges of using AI in finance is the "black box" nature of some machine learning models, which can make it difficult to understand how decisions are made. Regulators are increasingly demanding transparency in AI decision-making processes to ensure accountability.

3.3 Ethical Use of Data

Financial institutions must be mindful of data privacy and ethical considerations when using AI agents. Ensuring that customer data is handled responsibly and in compliance with data protection laws is crucial for maintaining trust and compliance.

4. Conclusion

AI agents are becoming indispensable tools in the finance industry, enabling faster and more accurate decision-making in trading, risk assessment, and fraud detection. However, as these technologies become more prevalent, it is essential for financial institutions to navigate the regulatory landscape carefully and adopt ethical practices to foster trust and accountability. The next chapter will explore the process of building conversational agents, highlighting design principles and tools for creating effective AI-driven customer interactions.

Chapter 15: Building Conversational Agents

Conversational agents, commonly referred to as chatbots or virtual assistants, are a pivotal component of AI technology that enhances user interaction across various platforms. These agents utilize natural language processing (NLP) and machine learning to understand, process, and respond to human language in a conversational manner. This chapter delves into the design principles for creating effective conversational agents, the tools available for their development, and the significance of user experience in ensuring successful interaction.

1. Design Principles for Chatbots and Virtual Assistants

1.1 Clarity and Purpose

The first principle in designing conversational agents is to define a clear purpose. Whether it's for customer service, information retrieval, or personal assistance, the agent should have a specific function that aligns with user needs. A well-defined purpose helps guide the design process and ensures that interactions are meaningful.

1.2 Natural Language Understanding

Effective conversational agents must excel in natural language understanding (NLU). This involves:

- **Intent Recognition:** Identifying the user's intention behind their input. For example, if a user types "What's the weather like today?", the intent is to get weather information.
- **Entity Recognition:** Extracting relevant information from the user's input, such as dates, locations, or names. For instance, in the query "Book a flight to New York on June 5th," "New York" is the destination, and "June 5th" is the date.

1.3 User-Centric Design

The design of conversational agents should prioritize the user experience. This includes:

- **Personalization:** Tailoring interactions based on user preferences, past interactions, and contextual information to create a more engaging experience.
- **Feedback Mechanism:** Implementing ways for users to provide feedback on their interactions. This can be through rating responses or flagging unhelpful answers, which can inform future improvements.
- **Simplicity:** Keeping the dialogue simple and intuitive. Avoid using jargon and complex language; instead, focus on clear and concise communication.

1.4 Error Handling

Conversational agents must be equipped to handle errors gracefully. This can include:

- **Clarifying Questions:** If the agent does not understand a user's input, it should ask clarifying questions to ensure comprehension.
- **Fallback Options:** Providing users with alternative actions or directing them to human representatives when the agent cannot fulfill their requests.

2. Tools for Developing Conversational AI

Building conversational agents involves leveraging various tools and platforms that facilitate the development and deployment of these technologies.

2.1 Natural Language Processing Frameworks

Several NLP frameworks are popular among developers for creating conversational agents:

- **Dialogflow:** A Google-owned platform that provides tools for building conversational interfaces. It includes features for intent recognition, entity extraction, and integrating with various messaging platforms.
- **Microsoft Bot Framework:** This comprehensive framework allows developers to create and manage intelligent bots. It supports multiple channels, including websites, Microsoft Teams, and Facebook Messenger.
- **Rasa:** An open-source machine learning framework designed for building contextual AI assistants. Rasa provides tools for training custom NLU models and managing dialogue effectively.

2.2 Development Platforms

In addition to NLP frameworks, development platforms can streamline the process:

- **Amazon Lex:** A service for building conversational interfaces using voice and text, which is part of the AWS suite. It integrates easily with other AWS services, enabling the development of sophisticated applications.
- **IBM Watson Assistant:** A powerful tool for creating conversational agents, Watson Assistant uses advanced NLP and machine learning techniques to understand user queries and respond appropriately.

2.3 Prototyping Tools

Prototyping tools enable designers to visualize and test the conversational flow before implementation:

- **Botmock:** This tool allows users to design and prototype conversational interfaces, providing a visual representation of dialogues and user interactions.
- **Voiceflow:** A platform that facilitates the design of voice apps and chatbots, Voiceflow allows users to create, test, and launch conversational agents visually.

3. User Experience and Conversational Design
3.1 Importance of User Experience

User experience (UX) is critical in the design of conversational agents. A positive UX leads to higher user satisfaction, increased engagement, and better retention rates. Designers should focus on creating a seamless interaction that feels natural and intuitive.

3.2 Conversational Design Best Practices

To ensure an optimal user experience, consider the following best practices:

- **Design for Context:** Tailor conversations based on the user's context, such as their location, previous interactions, and preferences. This can enhance relevance and engagement.
- **Use Personalization Wisely:** While personalization can improve user experience, it should be implemented cautiously to avoid overwhelming users with irrelevant information.
- **Iterate Based on Feedback:** Continuously improve the conversational agent by analyzing user interactions, gathering feedback, and making necessary adjustments.

4. Conclusion

Building effective conversational agents requires a thoughtful approach to design, leveraging advanced NLP technologies, and prioritizing user experience. As the technology evolves, these agents will play an increasingly vital role in facilitating interactions across various domains. In the next chapter, we will explore the integration of AI agents with the Internet of Things (IoT) and how this combination enhances various applications.

Chapter 16: Integrating AI Agents with IoT

The Internet of Things (IoT) represents a significant shift in how devices connect, communicate, and interact. As everyday objects become embedded with sensors, software, and connectivity, they generate vast amounts of data that can be leveraged to create intelligent systems. Integrating AI agents with IoT opens up new possibilities for automation, enhanced decision-making, and improved user experiences across various domains. This chapter explores the relationship between AI agents and IoT, detailing how AI enhances IoT applications and showcasing relevant case studies.

1. Understanding the Internet of Things (IoT)
1.1 Definition and Scope

The Internet of Things refers to the interconnection of physical devices, vehicles, buildings, and other objects embedded with sensors, software, and network connectivity that enables them to collect and exchange data. IoT devices can range from household items like smart thermostats and refrigerators to industrial equipment like machinery and sensors.

1.2 Key Components of IoT

IoT comprises several essential components:

- **Devices and Sensors:** Physical objects that collect data from their environment. Sensors can monitor temperature, humidity, light, motion, and more.
- **Connectivity:** The means through which devices communicate with each other and with cloud-based platforms. This can include Wi-Fi, Bluetooth, Zigbee, cellular networks, and more.
- **Data Processing and Analysis:** The gathered data is processed to extract valuable insights. This often occurs in the cloud, where AI algorithms analyze the data to provide actionable intelligence.
- **User Interface:** The platform through which users interact with IoT systems, often through mobile apps or web interfaces.

2. How AI Agents Enhance IoT Applications

2.1 Data Analysis and Decision-Making

AI agents significantly enhance the capabilities of IoT systems through advanced data analysis. By applying machine learning algorithms to the vast amounts of data generated by IoT devices, AI agents can:

- **Predictive Analytics:** Forecast trends and behaviors based on historical data. For example, predictive maintenance can alert manufacturers about potential equipment failures before they occur, allowing for timely interventions.
- **Anomaly Detection:** Identify unusual patterns in data that may indicate malfunctions or security threats. AI agents can analyze sensor data in real time, flagging deviations from expected behaviors.

2.2 Automation and Control

AI agents can automate responses based on data inputs from IoT devices. This automation can streamline processes across various industries, such as:

- **Smart Homes:** AI agents can control home automation systems, adjusting lighting, heating, and security based on user preferences and environmental conditions.
- **Smart Cities:** AI agents can manage traffic flow, optimize energy consumption in buildings, and enhance public safety by analyzing data from various city sensors.

2.3 Personalized Experiences

By integrating AI with IoT, organizations can create personalized experiences for users. AI agents can learn user preferences over time and adjust services accordingly. For instance:

- **Wearable Health Devices:** AI agents can analyze data from fitness trackers to provide personalized health recommendations, such as exercise plans or dietary advice based on the user's activity levels and health goals.
- **Retail Environments:** Smart shelves and IoT-connected devices can monitor customer behavior in real time, allowing AI agents to offer tailored promotions and product recommendations.

3. Case Studies of IoT and AI Agent Integration

3.1 Case Study: Smart Agriculture

In agriculture, IoT sensors collect data on soil moisture, temperature, and crop health. AI agents analyze this data to provide farmers with insights that enhance crop yield and resource management. By predicting the optimal times for planting and harvesting, farmers can make informed decisions that lead to increased productivity and sustainability.

3.2 Case Study: Healthcare Monitoring

IoT devices are used in remote patient monitoring systems to collect vital signs and health data. AI agents analyze this data in real time, alerting healthcare providers to potential health issues before they escalate. This integration allows for proactive patient care and reduces the need for hospital visits.

3.3 Case Study: Smart Transportation

In smart transportation systems, IoT sensors collect data on traffic patterns, vehicle locations, and road conditions. AI agents use this data to optimize traffic flow, reduce congestion, and enhance public transportation services. For instance, AI can predict traffic patterns and adjust traffic signals accordingly, improving overall efficiency.

4. Conclusion

Integrating AI agents with IoT transforms how industries operate, enabling enhanced data analysis, automation, and personalized experiences. As technology advances, the synergy between AI and IoT will continue to drive innovation and create smarter, more efficient systems across various sectors. In the next chapter, we will explore the role of AI agents in gaming and entertainment, examining how they contribute to immersive experiences and dynamic content creation.

Chapter 17: AI in Gaming and Entertainment

The integration of AI agents into gaming and entertainment has revolutionized how stories are told, how players interact with virtual worlds, and how immersive experiences are created. AI agents enhance gameplay by providing intelligent behaviors to non-playable characters (NPCs), generating content dynamically, and personalizing experiences based on player behavior. This chapter delves into the role of AI agents in game design, procedural content generation, NPC behavior, and explores future trends that will shape the gaming landscape.

1. Role of AI Agents in Game Design

1.1 Intelligent NPCs

One of the most significant contributions of AI agents in gaming is the development of intelligent NPCs. Traditionally, NPCs had limited interactions based on predefined scripts, which made their behavior predictable and often unengaging. AI agents, equipped with advanced algorithms, can exhibit more complex and realistic behaviors, enhancing player immersion. They can learn from player actions, adapt their strategies, and respond dynamically to changing game environments.

Example: The Last of Us Part II

In this critically acclaimed game, NPCs exhibit realistic behaviors. Enemies coordinate tactics, communicate with each other, and respond to the player's actions in real-time, creating a tense and engaging gameplay experience.

1.2 Enhanced Game Worlds

AI agents also contribute to creating expansive and vibrant game worlds. They can manage the behavior of various elements within a game environment, such as weather systems, wildlife, and population dynamics, making the world feel alive and responsive.

2. Procedural Content Generation
2.1 Definition and Importance

Procedural content generation (PCG) refers to the algorithmic creation of game content through automated processes rather than manual design. This approach can be used to generate levels, quests, characters, and even storylines. AI agents play a crucial role in PCG by ensuring that the content created is coherent, engaging, and tailored to player preferences.

2.2 Techniques in PCG

- **Random Generation:** Basic PCG often relies on randomization to create levels or encounters. While this can lead to unique experiences, it may lack coherence.
- **Rules-Based Systems:** More advanced PCG utilizes rules to generate content. For instance, a level might be created based on specific design principles, ensuring playability and challenge.
- **Machine Learning Approaches:** Some games employ machine learning algorithms to analyze player behavior and preferences. This data can guide the generation of content that aligns with what players find enjoyable.

Example: No Man's Sky

In *No Man's Sky*, an entire universe is procedurally generated, with billions of planets, each offering unique environments, creatures, and ecosystems. AI-driven algorithms create a vast array of content, ensuring that players have new experiences every time they explore.

3. Future Trends in Gaming with AI

3.1 Personalized Gaming Experiences

As AI continues to evolve, the potential for personalized gaming experiences increases. AI agents can analyze a player's style, preferences, and skill level to tailor challenges, storylines, and even NPC interactions, creating a uniquely engaging experience.

3.2 Adaptive Difficulty

Adaptive difficulty systems can adjust the game's challenges based on player performance. AI agents can monitor player skills and progress in real time, modifying the game's difficulty level to maintain a balance between challenge and enjoyment.

3.3 AI-Driven Storytelling

AI agents can dynamically alter narratives based on player choices and actions, creating branching storylines that react to player input. This capability could lead to more immersive storytelling experiences, where players genuinely feel the impact of their decisions.

3.4 Enhanced Player Interactions

With advancements in natural language processing (NLP), AI agents could enable players to engage in more realistic conversations with NPCs. Instead of scripted dialogues, players might have the ability to speak freely, leading to unique interactions and outcomes.

4. Conclusion

The integration of AI agents in gaming and entertainment is reshaping the landscape of interactive experiences. From creating intelligent NPCs to enabling procedural content generation and personalized gameplay, AI enhances engagement and immersion for players. As technology continues to advance, the future of gaming holds exciting possibilities, including adaptive storytelling, personalized challenges, and enriched player interactions. In the next chapter, we will explore the critical role of security in AI agents, examining the implications of AI in cybersecurity and how it can be utilized for threat detection and response.

Chapter 18: Security and AI Agents

As AI agents become increasingly integrated into various industries and everyday applications, their role in security has never been more critical. This chapter explores the cybersecurity implications of AI agents, their application in threat detection and response, and the challenges and solutions associated with securing AI systems. Understanding these facets is essential for developing resilient AI solutions that protect against malicious activities while leveraging the strengths of AI technology.

1. Cybersecurity Implications of AI Agents

1.1 The Dual Nature of AI

AI agents possess the ability to enhance security measures while simultaneously presenting new vulnerabilities. On one hand, they can improve threat detection capabilities through advanced algorithms and data analytics. On the other hand, they can also be exploited by cybercriminals to launch sophisticated attacks, creating a double-edged sword scenario.

1.2 AI in Cyber Threats

Cybercriminals are increasingly using AI to enhance their attack strategies. Automated bots can exploit vulnerabilities, perform phishing attacks at scale, or manipulate data systems. The sophistication of these attacks has grown, making it essential for organizations to adopt AI-driven security measures that can adapt to evolving threats.

2. AI in Threat Detection and Response

2.1 Enhanced Threat Detection

AI agents can analyze vast amounts of data in real-time, identifying patterns that may indicate security breaches. They leverage machine learning models to discern between normal and anomalous behavior, which is critical for detecting potential intrusions early.

- **Anomaly Detection:** By training on historical data, AI agents can establish a baseline of normal operations. Any deviation from this baseline can trigger alerts for further investigation.
- **Predictive Analytics:** AI systems can forecast potential threats by analyzing trends and behaviors, allowing organizations to implement preventative measures before incidents occur.

2.2 Automated Response

AI agents not only detect threats but can also respond automatically, mitigating potential damage. Automated incident response can involve isolating compromised systems, blocking unauthorized access, and deploying patches or updates to vulnerabilities.

- **Incident Response Systems:** AI-driven systems can evaluate the severity of threats and initiate predefined responses without human intervention, drastically reducing response times.
- **Continuous Monitoring:** AI agents can provide 24/7 surveillance, ensuring that threats are identified and addressed in real-time, further enhancing the security posture of organizations.

3. Challenges in Securing AI Systems

3.1 Data Privacy Concerns

The data used to train AI systems is often sensitive, raising concerns about privacy and compliance with regulations like GDPR. Organizations must ensure that they handle data responsibly, utilizing anonymization and encryption where necessary to protect user information.

3.2 Bias and Fairness

AI systems can inadvertently perpetuate bias, which can lead to unfair treatment of users or the mishandling of threats. Ensuring fairness and transparency in AI algorithms is crucial for maintaining trust in security systems.

3.3 Vulnerabilities in AI Models

AI models themselves can be vulnerable to attacks such as adversarial attacks, where malicious inputs are designed to deceive the AI into making incorrect predictions or classifications. Securing AI models against such attacks requires ongoing research and development.

4. Solutions to Enhance Security

4.1 Robust AI Development Practices

To secure AI systems, organizations should adopt best practices in AI development, including:

- **Regular Audits:** Conduct audits of AI models and algorithms to identify biases and vulnerabilities.
- **Transparency:** Ensure transparency in how AI agents make decisions, allowing for easier identification of potential security issues.
- **Collaboration:** Foster collaboration between AI developers and cybersecurity experts to create more resilient systems.

4.2 Multi-layered Security Approaches

A multi-layered security approach can significantly enhance the effectiveness of AI-driven security measures. This includes:

- **Network Security:** Employing firewalls, intrusion detection systems (IDS), and other security measures to protect network boundaries.
- **Endpoint Security:** Implementing security measures at the endpoints where AI agents operate, ensuring that they are not compromised.
- **User Training:** Educating users about security best practices and the potential threats posed by AI technology can help create a more secure environment.

5. Conclusion

The integration of AI agents into cybersecurity presents both opportunities and challenges. While AI has the potential to revolutionize threat detection and response, organizations must remain vigilant about the risks associated with AI technology. By adopting robust security practices, fostering collaboration between AI and cybersecurity experts, and employing multi-layered security strategies, organizations can harness the power of AI agents while protecting against emerging threats. In the next chapter, we will explore the future of AI agents, examining emerging trends and predictions that will shape the evolution of this technology in the coming years.

Chapter 19: The Future of AI Agents

As we navigate through the 21st century, the landscape of artificial intelligence (AI) is evolving rapidly. AI agents, in particular, are at the forefront of this transformation, influencing various sectors and redefining the relationship between technology and human society. In this chapter, we will explore emerging trends in AI technologies, predict the future evolution of AI agents, and examine their potential impact on society and jobs.

1. Emerging Trends in AI Technologies

1.1 Enhanced Machine Learning Algorithms

The development of more sophisticated machine learning algorithms is a critical trend shaping the future of AI agents. Techniques like deep learning, ensemble methods, and transfer learning are becoming more refined, enabling AI agents to process vast amounts of data more efficiently and make better predictions. These advancements are particularly significant in fields such as healthcare, finance, and autonomous systems, where the ability to analyze complex datasets in real time can lead to improved outcomes.

1.2 Natural Language Understanding (NLU)

Natural Language Processing (NLP) is advancing beyond mere language generation to encompass Natural Language Understanding (NLU). This evolution allows AI agents to comprehend context, tone, and intent, leading to more meaningful interactions in conversational AI applications. Enhanced NLU capabilities will make chatbots and virtual assistants more effective, fostering deeper engagement with users and transforming customer service experiences.

1.3 Integration of AI with IoT

The convergence of AI with the Internet of Things (IoT) is set to redefine how we interact with our environment. AI agents integrated into IoT devices can process and analyze data locally, making real-time decisions that enhance efficiency and responsiveness. This integration is vital for smart cities, home automation, and industrial applications, where timely insights and actions can significantly improve operational effectiveness.

1.4 Ethical AI Development

As the capabilities of AI agents expand, so does the importance of ethical considerations in their development. Organizations are increasingly prioritizing fairness, transparency, and accountability in AI systems. The future will likely see the emergence of standardized ethical frameworks and best practices for AI development, ensuring that AI agents operate within guidelines that prioritize human rights and social good.

2. Predictions for the Evolution of AI Agents

2.1 More Autonomous Systems

We can expect AI agents to evolve into more autonomous systems capable of independent decision-making across various applications. From autonomous vehicles to drone delivery systems, these agents will leverage advanced AI technologies to navigate complex environments and make real-time choices without human intervention. As autonomy increases, so too will the need for robust safety and regulatory measures.

2.2 Personalization at Scale

AI agents will continue to enhance personalization across platforms, tailoring experiences to individual users based on their preferences and behaviors. This trend will be particularly evident in marketing, entertainment, and e-commerce, where AI-driven recommendations can significantly enhance user engagement and satisfaction.

2.3 Increased Collaboration Between AI and Humans

The relationship between humans and AI agents will shift from one of simple interaction to collaboration. AI agents will not only support human decision-making but will also work alongside humans to augment their capabilities. This collaborative dynamic will likely lead to the emergence of new roles and skill sets, emphasizing the need for individuals to adapt and evolve alongside AI technology.

3. Impact of AI Agents on Society and Jobs

3.1 Workforce Transformation

The integration of AI agents into various sectors will inevitably lead to workforce transformation. While AI will automate repetitive tasks, it will also create new job opportunities in AI development, maintenance, and oversight. Workers will need to develop new skills to thrive in an increasingly AI-driven job market, emphasizing the importance of continuous education and training.

3.2 Social Good Initiatives

AI agents have the potential to address pressing societal challenges. From healthcare innovations that improve patient outcomes to environmental monitoring systems that combat climate change, the applications of AI agents in social good initiatives are vast. As organizations recognize the value of leveraging AI for positive societal impact, we can expect to see more collaborative efforts focused on harnessing AI for the greater good.

3.3 Ethical and Regulatory Considerations

As AI agents become more integrated into society, ethical and regulatory considerations will become paramount. Policymakers will need to develop frameworks that govern the use of AI, ensuring that these technologies are deployed responsibly. This will involve addressing issues such as data privacy, algorithmic bias, and accountability, ultimately shaping the ethical landscape in which AI agents operate.

4. Conclusion

The future of AI agents is bright, with advancements in technology poised to reshape industries, enhance human capabilities, and address global challenges. As we move forward, it is crucial to remain mindful of the ethical considerations and societal implications of these technologies. By fostering collaboration between AI and humans and prioritizing responsible development, we can harness the full potential of AI agents while ensuring a positive impact on society. In the next chapter, we will explore the dynamics of human-AI collaboration, focusing on how these two forces can work together to enhance productivity and innovation.

Chapter 20: Human-AI Collaboration

As artificial intelligence (AI) continues to evolve, the interplay between humans and AI agents is becoming increasingly important. This chapter explores the dynamics of collaboration between humans and AI agents, focusing on how these partnerships enhance human capabilities, improve productivity, and shape the future of work.

1. Understanding Collaboration Between Humans and AI Agents
1.1 The Nature of Collaboration

Collaboration between humans and AI agents is not merely about one entity performing tasks for the other. Instead, it is about leveraging the strengths of both to achieve common goals. Humans possess emotional intelligence, creativity, and contextual understanding, while AI agents excel at processing vast amounts of data quickly, executing repetitive tasks, and identifying patterns. This complementary relationship allows for enhanced decision-making and more efficient processes.

1.2 Levels of Collaboration

The collaboration between humans and AI can be classified into several levels, each with distinct characteristics:

- **Assistance**: AI agents perform specific tasks that support human activities. For example, AI tools can analyze data to provide insights that inform human decisions, such as in financial forecasting.
- **Augmentation**: AI agents enhance human abilities by providing additional insights or capabilities. For instance, in healthcare, AI can assist doctors by analyzing medical images, allowing for faster and more accurate diagnoses.
- **Autonomous Collaboration**: In some cases, AI agents can operate independently while working towards a shared objective with human input. This is common in manufacturing, where AI-controlled robots collaborate with human workers on assembly lines.

2. Enhancing Human Capabilities with AI

2.1 Decision-Making Support

AI agents can analyze large datasets far beyond human capacity, extracting actionable insights that inform decision-making. For example, AI-driven analytics platforms can identify market trends or customer behaviors, enabling businesses to make data-informed strategic decisions.

2.2 Automation of Repetitive Tasks

By automating repetitive and mundane tasks, AI agents free up human resources to focus on more complex and creative activities. For instance, AI can handle data entry, scheduling, and basic customer inquiries, allowing employees to concentrate on higher-level problem-solving and innovation.

2.3 Improving Creativity

AI tools can enhance creative processes by offering suggestions, generating ideas, or even creating original content. In the arts, AI can assist musicians in composing music or help writers brainstorm story ideas, fostering a collaborative creative environment.

3. The Future of Work with AI Agents

3.1 Evolving Job Roles

As AI agents become more integrated into workplaces, job roles will inevitably evolve. Workers will need to adapt by developing new skills that complement AI technologies. Skills such as emotional intelligence, critical thinking, and creativity will become increasingly valuable in a world where routine tasks are automated.

3.2 Collaboration Skills

The ability to collaborate effectively with AI will be essential. Workers must learn how to interpret AI-generated insights, communicate with AI systems, and integrate AI tools into their workflows. Training programs focusing on human-AI collaboration will be crucial in preparing the workforce for these changes.

3.3 Ethical Considerations

With increased reliance on AI agents, ethical considerations must guide human-AI collaboration. Transparency, accountability, and fairness are essential principles that should be upheld to ensure that AI systems are used responsibly. Organizations must develop ethical frameworks to govern AI use, ensuring that human oversight is maintained and that biases are addressed.

4. Conclusion

The collaboration between humans and AI agents presents an exciting opportunity to enhance productivity, creativity, and decision-making. As we look to the future, it is essential to foster a culture of collaboration that values both human ingenuity and AI capabilities. By embracing this partnership, we can navigate the challenges and opportunities of an AI-driven world, paving the way for a more productive and innovative future.

In the next chapter, we will examine case studies of successful AI agent implementations across various industries, highlighting lessons learned and best practices that can inform future initiatives.

Chapter 21: Case Studies in AI Agent Implementation

In this chapter, we will explore several case studies that illustrate successful implementations of AI agents across various industries. These examples highlight the practical applications of AI technology, the challenges faced during development, and the lessons learned. By examining these cases, we can glean insights that will inform future AI projects and showcase best practices.

1. AI in Healthcare: IBM Watson for Oncology
1.1 Overview

IBM Watson for Oncology is an AI system designed to assist healthcare professionals in diagnosing and recommending treatment options for cancer patients. It utilizes natural language processing (NLP) and machine learning to analyze vast amounts of medical literature, clinical trial data, and patient records.

1.2 Implementation

In a pilot program at the Manipal Comprehensive Cancer Center in India, Watson analyzed data from over 300 cancer patients to generate personalized treatment plans. The AI system was trained on a comprehensive database of oncology knowledge, enabling it to offer evidence-based treatment recommendations.

1.3 Lessons Learned

- **Data Quality**: The success of AI systems depends heavily on the quality of the data fed into them. Ensuring that the training data is comprehensive, up-to-date, and free from biases is crucial.
- **Collaboration with Experts**: The integration of AI in healthcare requires close collaboration between AI developers and medical professionals to ensure that the system meets clinical needs and adheres to ethical standards.

2. AI in Finance: Betterment's Robo-Advisory Platform

2.1 Overview

Betterment is a robo-advisor that uses AI algorithms to provide personalized investment advice and portfolio management services. By leveraging machine learning, Betterment aims to optimize investment strategies based on individual customer profiles and market conditions.

2.2 Implementation

Betterment's platform collects user data through surveys and analyzes it to create tailored investment plans. The AI algorithms continuously monitor the market and adjust portfolios based on changing economic conditions, user goals, and risk tolerances.

2.3 Lessons Learned

- **User Experience**: Simplifying the user interface and ensuring clear communication about how the AI system operates enhances user trust and satisfaction.
- **Regulatory Compliance**: Financial services must navigate complex regulatory landscapes, necessitating AI systems to be designed with compliance in mind from the outset.

3. AI in Retail: Sephora Virtual Artist
3.1 Overview

Sephora, a global cosmetics retailer, implemented the Virtual Artist app to enhance customer engagement. This AI-powered tool uses augmented reality (AR) and computer vision to allow users to try on different makeup products virtually.

3.2 Implementation

By uploading a selfie, users can see how various products look on their skin in real-time. The app leverages facial recognition technology to accurately overlay products, providing a personalized shopping experience.

3.3 Lessons Learned

- **Innovative Marketing**: Combining AI with AR creates unique marketing opportunities that can attract tech-savvy consumers.
- **Feedback Loops**: Incorporating user feedback into the app's development cycle can lead to continuous improvement and higher user satisfaction.

4. AI in Manufacturing: Siemens' Predictive Maintenance

4.1 Overview

Siemens has integrated AI agents into its manufacturing processes for predictive maintenance, which helps anticipate equipment failures before they occur. This approach minimizes downtime and optimizes operational efficiency.

4.2 Implementation

Using IoT sensors, Siemens collects data from machinery and applies machine learning algorithms to identify patterns indicative of potential failures. This data-driven approach allows maintenance teams to act proactively, reducing the likelihood of unexpected breakdowns.

4.3 Lessons Learned

- **Data Integration**: Seamlessly integrating data from various sources (sensors, machines, and maintenance logs) is vital for the effectiveness of predictive maintenance systems.
- **Cost Savings**: The long-term cost savings associated with reduced downtime and maintenance expenses can justify the initial investment in AI technologies.

5. AI in Customer Service: LivePerson Chatbots

5.1 Overview

LivePerson provides AI-powered chatbots that enhance customer service experiences for businesses. These chatbots engage with customers in real time, answering questions and guiding them through the purchasing process.

5.2 Implementation

By integrating natural language processing and machine learning, LivePerson's chatbots can understand and respond to customer inquiries effectively. The system is designed to hand off complex queries to human agents when necessary, ensuring high-quality customer service.

5.3 Lessons Learned

- **Continuous Learning**: AI chatbots improve over time through continuous learning from customer interactions, leading to enhanced accuracy and customer satisfaction.
- **Human Touch**: Striking the right balance between automation and human interaction is essential to maintaining customer trust and satisfaction.

Conclusion

The case studies highlighted in this chapter demonstrate the diverse applications of AI agents across industries and their potential to drive innovation, enhance efficiency, and improve user experiences. While each implementation faces unique challenges, the lessons learned provide valuable insights that can guide future AI projects. As organizations continue to explore AI technologies, understanding these case studies will help in designing more effective and responsible AI agents.

In the next chapter, we will discuss the common challenges faced in AI development, offering strategies for overcoming these obstacles to achieve successful AI agent implementation.

Chapter 22: Overcoming Challenges in AI Development

The development of AI agents presents unique opportunities and challenges across various domains. As organizations seek to harness the power of AI, understanding these challenges is crucial for successful implementation. This chapter will explore common obstacles in developing AI agents, provide strategies for effective project management, and emphasize the importance of interdisciplinary collaboration.

1. Common Obstacles in Developing AI Agents

1.1 Data Quality and Availability

AI agents rely heavily on data for training and operation. Poor-quality or insufficient data can lead to ineffective models that do not meet user needs. Organizations often struggle to collect clean, labeled, and relevant datasets, impacting the performance of their AI systems.

1.2 Integration with Existing Systems

Integrating AI agents into existing workflows and systems can be a significant challenge. Legacy systems may not be compatible with new AI technologies, requiring additional resources and expertise for seamless integration.

1.3 Algorithmic Bias

Bias in AI systems can arise from biased training data or flawed algorithms, leading to unfair or discriminatory outcomes. Ensuring fairness and transparency in AI models is critical to maintaining trust and compliance with ethical standards.

1.4 Technical Complexity

Developing AI agents involves various technologies, methodologies, and frameworks, which can overwhelm teams without the necessary expertise. The complexity of machine learning, natural language processing, and other AI techniques requires specialized knowledge that may not always be readily available.

1.5 Regulatory and Compliance Issues

As AI technologies evolve, regulatory frameworks are also developing. Organizations must navigate these regulations, ensuring compliance while pursuing innovation. This can lead to uncertainty and caution in adopting new AI solutions.

1.6 Change Management

Implementing AI agents often requires changes in organizational processes and employee roles. Resistance to change can hinder adoption, necessitating effective change management strategies to facilitate a smooth transition.

2. Strategies for Effective Project Management

2.1 Define Clear Objectives

Establishing clear goals for the AI project is crucial. Teams should define what success looks like and the specific problems the AI agent aims to solve. This clarity helps align stakeholders and provides a framework for measuring progress.

2.2 Foster Interdisciplinary Collaboration

AI projects often require input from various disciplines, including data science, software engineering, domain expertise, and user experience design. Encouraging collaboration between these disciplines can enhance the quality of the AI agent and address potential challenges early on.

2.3 Implement Agile Methodologies

Adopting agile project management methodologies can help teams iterate quickly and respond to changing requirements. Agile practices promote flexibility, allowing teams to adapt their approach based on user feedback and new insights.

2.4 Prioritize Data Governance

Implementing robust data governance practices ensures data quality, security, and compliance. Organizations should establish clear protocols for data collection, storage, and usage to mitigate risks associated with data-related issues.

2.5 Invest in Training and Development

Providing ongoing training for team members on AI technologies and best practices is essential for building expertise and confidence. Continuous learning opportunities can help teams stay updated with industry trends and overcome technical challenges.

3. Importance of Interdisciplinary Collaboration

3.1 Bridging Knowledge Gaps

Interdisciplinary collaboration helps bridge knowledge gaps between technical and non-technical teams. Data scientists, engineers, and business stakeholders can work together to align AI development with organizational goals and user needs.

3.2 Enhancing Innovation

Diverse perspectives foster innovation by encouraging creative problem-solving and the exploration of new ideas. Collaboration can lead to novel approaches to challenges and the development of more effective AI agents.

3.3 Ensuring Ethical Considerations

Bringing together experts from various fields, including ethics, law, and social sciences, helps ensure that AI development considers broader societal implications. This holistic approach can help mitigate risks associated with bias and unethical practices.

Conclusion

Overcoming the challenges in developing AI agents requires a comprehensive approach that emphasizes collaboration, effective project management, and a focus on data quality. By recognizing and addressing these obstacles, organizations can enhance their ability to implement successful AI solutions that meet user needs and drive innovation. In the next chapter, we will explore valuable learning resources for mastering AI agents, equipping you with the knowledge and skills needed to excel in this rapidly evolving field.

Chapter 23: Learning Resources for Mastering AI Agents

As the field of artificial intelligence and AI agents continues to evolve rapidly, having access to the right resources is crucial for both newcomers and seasoned professionals. This chapter provides a curated list of books, online courses, websites, and communities that will help you deepen your understanding of AI agents, develop essential skills, and stay updated on industry trends.

1. Recommended Books

1.1 General AI and Machine Learning

- **"Artificial Intelligence: A Guide to Intelligent Systems" by Michael Negnevitsky**
  This book provides a solid foundation in AI concepts, including intelligent systems, expert systems, and neural networks.

- **"Hands-On Machine Learning with Scikit-Learn, Keras, and TensorFlow" by Aurélien Géron**
  A practical guide to implementing machine learning algorithms using popular Python libraries, ideal for hands-on learning.

1.2 Natural Language Processing

- **"Speech and Language Processing" by Daniel Jurafsky and James H. Martin**
  A comprehensive textbook on NLP that covers the theory and practical applications of language processing technologies.
- **"Natural Language Processing with Python" by Steven Bird, Ewan Klein, and Edward Loper**
  A hands-on guide for working with text data in Python, using the Natural Language Toolkit (NLTK).

1.3 Reinforcement Learning

"Reinforcement Learning: An Introduction" by Richard S. Sutton and Andrew G. Barto

1.4 AI Ethics and Policy

"Weapons of Math Destruction: How Big Data Increases Inequality and Threatens Democracy" by Cathy O'Neil

2. Online Courses
2.1 Platforms Offering AI Courses

- **Coursera**

    Offers a wide range of courses in AI, including "AI For Everyone" by Andrew Ng and specialized tracks in machine learning and deep learning.

- **edX**

    Provides courses from top universities on AI topics, including MIT's "Introduction to Computer Science and Programming Using Python" and Harvard's "Data Science Professional Certificate."

- **Udacity**

    Known for its nanodegree programs, Udacity offers specialized courses like "AI Programming with Python" and "Deep Learning."

2.2 Specialized Courses

- **Deep Learning Specialization by Andrew Ng (Coursera)**

    A series of five courses that delve into deep learning techniques, ideal for those looking to specialize in this area.

- **Natural Language Processing Specialization (Coursera)**

    A focused curriculum on NLP that covers text processing, sentiment analysis, and more.

3. Websites and Blogs

3.1 Knowledge Hubs

- **Towards Data Science (Medium)**

  A popular platform for articles on data science and AI, featuring tutorials, case studies, and insights from practitioners.

- **KDnuggets**

  A leading site on AI and data science, offering articles, newsletters, and resources for professionals in the field.

3.2 Research and Publications

- **arXiv.org**

  A repository for research papers in various fields, including AI, machine learning, and NLP. It's a great resource for staying updated on the latest research.

- **Google AI Blog**

  Google's official blog on AI developments, sharing research findings and applications of AI technologies.

4. Building a Community

4.1 Online Forums and Groups

- **Stack Overflow**

  A valuable resource for troubleshooting coding issues related to AI and machine learning.

- **Reddit**

  Subreddits like r/MachineLearning and r/artificial provide discussions, news, and community support for AI enthusiasts.

4.2 Local Meetups and Conferences

- **Meetup.com**

    Look for local AI or machine learning groups where you can network with professionals and attend workshops.

- **Conferences**

    Major AI conferences such as NeurIPS, ICML, and AAAI are great opportunities to learn from leading experts and network with other professionals.

5. Staying Updated with Industry Trends

- **Newsletters**

    Subscribe to newsletters like "The AI Report" or "Import AI" for regular updates on AI trends and innovations.

- **Podcasts**

    Listening to AI-focused podcasts, such as "The TWIML AI Podcast" and "AI Alignment Podcast," can help you stay informed about the latest developments and discussions in the field.

Conclusion

Mastering AI agents requires continuous learning and adaptation to new technologies and methodologies. By utilizing the recommended resources in this chapter, you can build a solid foundation in AI, improve your skills, and stay engaged with the evolving landscape of AI agents. In the next chapter, we will explore the global impact of AI agents, particularly in social good initiatives and how they can address pressing global challenges.

Chapter 24: AI Agents and Global Impact

AI agents are increasingly becoming integral to addressing some of the most pressing challenges faced by societies worldwide. From enhancing healthcare delivery to improving educational access and addressing climate change, AI agents possess the capability to effect meaningful change in various domains. This chapter will explore the role of AI agents in social good initiatives, how they address global challenges, and present case studies of impactful AI projects.

1. AI Agents in Social Good Initiatives

AI agents are deployed across multiple sectors to drive social good, enhancing the quality of life and promoting sustainability. Here are some areas where AI agents are making a significant impact:

1.1 Healthcare Access and Improvement

AI agents are used to improve healthcare delivery, especially in underserved areas. They can facilitate remote patient monitoring, telemedicine, and personalized treatment plans, ensuring that individuals have access to necessary healthcare services.

Example:

Babylon Health

1.2 Education and Skill Development

AI agents in education facilitate personalized learning experiences, adapting to individual student needs. They can assist teachers in managing classrooms, grading assignments, and providing additional resources to students.

Example:

Knewton

1.3 Disaster Response and Management

AI agents play a crucial role in disaster preparedness and response. They analyze data from various sources to predict natural disasters, optimize response strategies, and enhance resource allocation.

Example:

Zebra Medical Vision

2. Addressing Global Challenges with AI Technology

AI agents can contribute solutions to a variety of global challenges, including climate change, poverty alleviation, and food security.

2.1 Climate Change Mitigation

AI agents help monitor environmental changes and optimize resource usage to combat climate change. They can analyze vast amounts of data from satellites, sensors, and IoT devices to provide insights that guide decision-making.

Example:

Climate AI

2.2 Poverty Alleviation

AI agents assist in identifying at-risk populations and designing targeted interventions. By analyzing socioeconomic data, they can help governments and NGOs allocate resources effectively.

Example:

GiveDirectly

2.3 Food Security and Agriculture

In agriculture, AI agents enhance crop management, predict yields, and optimize supply chains. They help farmers make informed decisions that increase productivity and sustainability.

Example:

Plantix

3. Case Studies of Impactful AI Projects
3.1 AI for Global Health
The AI4Health Initiative

3.2 AI in Wildlife Conservation
WildTrack

3.3 Smart Cities
IBM's Intelligent Operations Center

4. Future Opportunities for AI Agents in Social Good

As AI technology continues to advance, the potential for AI agents to tackle global challenges will only grow. Key areas of opportunity include:

- **Enhancing Predictive Analytics**: Improving models to anticipate social issues such as homelessness or unemployment through data analysis, allowing for proactive solutions.
- **Promoting Inclusivity**: Developing AI agents that are accessible to diverse populations, including those with disabilities, to ensure everyone benefits from technological advancements.
- **Fostering Collaboration**: Encouraging partnerships between AI developers, governments, and NGOs to create impactful solutions that leverage AI for the greater good.

Conclusion

AI agents hold immense potential to create positive societal impacts, addressing global challenges and promoting sustainable development. As we continue to explore and harness their capabilities, it is essential to approach their implementation responsibly and inclusively. In the concluding chapter, we will recap the key insights from this book and encourage a commitment to ethical AI practices and social responsibility.

Chapter 25: Conclusion and Call to Action

As we conclude this exploration of AI agents, it's essential to reflect on the transformative potential they hold across various domains. From enhancing business operations to solving global challenges, AI agents are not just tools but partners in innovation and efficiency. This chapter summarizes the key insights discussed throughout the book and presents a call to action for stakeholders to engage responsibly with AI technologies.

1. Key Insights from the Book

1.1 Understanding AI Agents

AI agents, defined by their ability to perceive, reason, and act within environments, are rapidly evolving. They range from simple reactive agents to complex autonomous systems, each designed for specific tasks and applications. Recognizing the diversity of AI agents is crucial for leveraging their capabilities effectively.

1.2 The Role of AI Technologies

The integration of various AI technologies—machine learning, natural language processing, computer vision, and more—forms the backbone of modern AI agents. Understanding these core technologies and their interdependencies enables developers and organizations to create more robust and capable agents.

1.3 Ethical and Responsible AI Development

Ethics must remain at the forefront of AI development. Issues such as bias, fairness, transparency, and accountability are not just theoretical considerations; they have real-world implications that can affect individuals and communities. Establishing frameworks for ethical AI is essential for fostering trust and ensuring that AI agents serve the common good.

1.4 Collaborative Human-AI Ecosystems

The future of work lies in collaboration between humans and AI agents. By enhancing human capabilities rather than replacing them, AI agents can lead to more productive and satisfying work environments. This symbiotic relationship should be nurtured through thoughtful design and implementation.

1.5 Global Impact and Social Good

AI agents are proving to be invaluable in addressing some of the world's most pressing issues, including healthcare access, education, and climate change. Their ability to analyze data and provide insights can drive meaningful social change, making it imperative that we continue to innovate in this space.

2. A Call to Action

2.1 For Developers and Researchers

- **Innovate Responsibly**: Prioritize ethical considerations in your AI projects. Strive to create AI agents that are transparent, fair, and beneficial to society.
- **Stay Informed**: The field of AI is continuously evolving. Keep up with the latest research, technologies, and best practices to ensure your skills and knowledge remain relevant.

2.2 For Businesses

- **Adopt AI Thoughtfully**: Assess the potential impact of AI agents on your operations. Implement them in ways that enhance employee productivity and customer experience while considering ethical implications.
- **Invest in Training**: Provide your workforce with training and resources to understand and collaborate with AI agents effectively. This will help create a culture of innovation and adaptability.

2.3 For Policymakers

- **Establish Guidelines**: Develop regulatory frameworks that promote the ethical development and deployment of AI agents. Ensure these guidelines foster innovation while protecting individuals' rights and societal interests.
- **Encourage Collaboration**: Facilitate partnerships between industry, academia, and government to advance AI research and address ethical concerns collaboratively.

2.4 For Society at Large

- **Engage in Dialogue**: Participate in discussions about AI and its implications for society. Raise awareness about the benefits and challenges posed by AI agents, fostering informed public discourse.
- **Advocate for Inclusivity**: Ensure that AI technologies are accessible and beneficial to all, particularly marginalized communities. Advocate for diverse representation in AI development to address potential biases and inequities.

3. Vision for the Future of AI Agents

As we look to the future, the potential of AI agents is vast. Emerging trends, such as the integration of AI with quantum computing and advancements in general artificial intelligence, promise to unlock new capabilities and applications. However, realizing this potential requires a collective commitment to ethical practices, innovation, and societal benefit.

The journey of mastering AI agents is not merely about technology but also about humanity's choices. By approaching this field with responsibility and foresight, we can harness the power of AI to create a better, more equitable world for everyone.

Closing Thoughts

Mastering AI agents is an ongoing endeavor that calls for continuous learning, collaboration, and ethical consideration. As we move forward, let us embrace the possibilities that AI offers while remaining steadfast in our commitment to using technology for the greater good. Together, we can shape a future where AI agents enhance lives, empower individuals, and drive positive change across the globe.

www.ingramcontent.com/pod-product-compliance
Lightning Source LLC
Chambersburg PA
CBHW082109220526

45472CB00009B/2114